Implementation and Verification of Distributed Control Systems

Dissertation

zur Erlangung des akademischen Grades
Doktoringenieur (Dr.-Ing.)

Genehmigt durch das

Zentrum für Ingenieurwissenschaften
der Martin-Luther-Universität Halle-Wittenberg
als organisatorische Grundeinheit für Forschung und Lehre im Range einer Fakultät
(§75 Abs. 1 HSG LSA, §1 Abs. 1 Grundordnung)

von

Herrn Dipl.-Ing. Christian Gerber
geboren am 17. Juli 1980 in Halle / Saale

Geschäftsführender Direktor (Dekan): Prof. Dr.-Ing. habil. Dr. h. c. Holm Altenbach

Gutachter:

1. Prof. Dr.-Ing. Hans-Michael Hanisch
2. Prof. Ph.D. Mengchu Zhou

Halle (Saale), den 28. März 2011

Reihe: Hallenser Schriften zur Automatisierungstechnik
herausgegeben von:
Prof. Dr. Hans-Michael Hanisch
Lehrstuhl für Automatisierungstechnik
Martin-Luther-Universität Halle-Wittenberg
Kurt-Mothes-Str. 1
06120 Halle/Saale

email: Hans-Michael.Hanisch@informatik.uni-halle.de

Bibliografische Information der Deutschen Nationalbibliothek

Die Deutsche Nationalbibliothek verzeichnet diese Publikation in der Deutschen Nationalbibliografie; detaillierte bibliografische Daten sind im Internet über http://dnb.d-nb.de abrufbar.

(Hallenser Schriften zur Automatisierungstechnik; 7)

ISBN 978-3-8325-2849-2

Logos Verlag Berlin GmbH
Comeniushof, Gubener Str. 47,
10243 Berlin
Tel.: +49 030 42 85 10 90
Fax: +49 030 42 85 10 92
INTERNET: http://www.logos-verlag.de

Vorwort

Die vorliegende Arbeit entstand während meiner wissenschaftlichen Tätigkeit am Lehrstuhl für Automatisierungstechnik an der Martin-Luther-Universität Halle-Wittenberg. Wie der Titel bereits suggeriert beschäftigte ich mich in unterschiedlichen Forschungsprojekten mit der Implementierung und der Verifikation Verteilter Steuerungssysteme. Hervorzuheben ist dabei das Verbundprojekt „Energieautarke Aktor- und Sensorsysteme (EnAS)“ über welches ich die Forschungsanlage zur Verfügung gestellt bekommen habe, und meine Theorien praktisch anwenden konnte.

Als erstes möchte ich an dieser Stelle meinem Doktorvater Prof. Dr. Hans-Michael Hanisch danken, der mich nach einer in der Industrie geschriebenen Diplomarbeit zurück in die Wissenschaft holte, und mir durch eine feste Anstellung als wissenschaftlicher Mitarbeiter die nötigen Mittel und Freiheiten gab, um diese Arbeit erfolgreich durchzuführen. Während der fast fünfjährigen Arbeit am Lehrstuhl ließ sich Hans-Michael Hanisch oft für verschiedene neue Ideen begeistern, und sorgte jedoch in den sich daraus ergebenden Diskussionen mit seiner kritischen Hinterfragung des Sinns und des Zweckes stets dafür, dass ich mich auf meine Arbeit fokussiere und nicht versuche alles auf einmal zu lösen.

Weiterhin möchte ich mich bei meinem zweiten Gutachter Prof. Ph.D. Mengchu Zhou bedanken, der mit seinen Anmerkung für eine Vervollständigung der Literaturliste sorgte und den englischen Ausdruck signifikant verbesserte.

Danken möchte ich auch Prof. Dr. Victor Dubinin, der mit seiner ersten Implementierung für $_{s}NCES$ den Ausschlag dafür gab, die in der Arbeit erstellte TNCES-Workbench mittels SWI-Prolog zu implementieren, ohne die die Durchführung der praktischen Arbeit undenkbar gewesen wäre.

Ferner möchte ich mich bei Dr. Valeriy Vyatkin bedanken, der mir durch seine Vorlesung ein erstes Verständnis für Verteilte System vermittelte und während seines Forschungsaufenthaltes am Lehrstuhl das Scheduling von Function Blocks hinterfragte und Anregungen zur Verbesserung gab.

Zu vergessen ist auch nicht Dr. Alois Zoitl, der mir unzählige Fragen zur Bedienung von 4DIAC und der Cross-Compilierung der FORTE für den Wago-IPC beantwortete, sowie Vorarbeiten zur Entwicklung der nötigen SIFB bereitstellte.

Natürlich sollen die verschiedenen Reviewer meiner Veröffentlichung und Sessions Chairs nicht vergessen werden, die mit Kritik und Anmerkung nicht gespart haben und so die Ausrichtung und wissenschaftliche Einordnung der Arbeit verbesserten.

Weiterhin möchte ich mich bei meinen Freunden bedanken und insbesonder bei den Rugbyjungs für den notwendigen Ausgleich und Bodenhaftung und mögen wir auch in Zukunft den nötigen Spielwitz an den Tag legen.

Ein abschließendes großes Dankeschön an meine Familie ohne deren Rückhalt und Ermutigungen gerade in der Endphase dem Schreiben dieses Buch undenkbar gewesen wäre.

Christian
im März 2011

Contents

List of Figures

Listings

List of Tables

Chapter 1

Introduction and Motivation

For medium-sized companies, it will become more and more important to create automation solutions for manufacturing plants, that are optimally coordinated with their customers. Thereby, these medium-sized companies can feature themselves by manufacturer independence and usage of hardware the customer prefers. Also the operating company of the manufacturing plant can feature itself by reacting fast and flexible to changed customer demands. Therefore, it has to be possible to refit an existing production line with new or other components, to change the production process as well as to bring new production lines faster into service. Consequently, there exist some key issues to modern industrial automation as

modularity to get a strict correspondence between control objects and the used mechatronic components [MFC99],

reusability of once developed control objects if the same mechatronic component occurs in another plant configuration,

portability of once developed control objects between several engineering environments [SZC+06],

flexibility to adapt existing production lines to new production processes as quickly as possible [SRE+08b],

extendibility and reconfigurability to prepare idle parts of the production line for the next planed job while the current product is still in production [HSZ+06].

This will ensure the competitiveness of the manufacturer in today's global market, but this increases also the complexity of the used distributed control system.
An appropriate way to resolve these issues is the object-oriented control implementation, which has been quite common since the late 90s. Thereby, the control function is encapsulated inside objects and the control engineer interconnects objects instead of implementing the system from scratch. To support the issue of reusability, the paradigm of the strict correspondence of each control object to a real mechatronic component is often used and the structure of the software objects is derived from the hierarchical structure of the mechanical components [VKP05]. Thus, it is suitable to rapidly implement and reconfigure an even complex manufacturing system using an object-oriented approach [BVT+08].

1.1 Focused on IEC 61499

Due to the fact that the approach of object-oriented control implementation comes from the application of software development, it can be realised with any kind of high level programming language, which would lead to various kinds of data exchange and synchronisation mechanism differing from a manufacturing system to another. Therefore, the Technical Committee 65 of the International Electronical Commission launched several standards as [IEC-61131-1] and [IEC-61499-1] to define the execution order and the mechanism of data exchange as well as the graphical representation. In the view of the author the main advantage of IEC 61499 in contrast to the prior published IEC 61131 is the system view with programming of control applications and mapping parts to one or more control devices, instead of programming each control device separately. Another advantage is the portability of once developed function blocks and applications due to the defined requirements for software tools at the second part of this standard [IEC-61499-2]. Thus, each control engineer can use the development environment he prefers. After the development of the distributed control system any other development environment can be used to update the system configuration. As reported in [GHE08] it is possible to use the available computing power and communication bandwidth better by executing only algorithms with changed data inputs and transmitting only data packages with changed data values, due to the new event-driven execution behaviour.
To benefit from this possibility the control engineer has to interconnect the function blocks carefully, because as described in [FV04] there exist critical function block interconnections as *event splitting* and *feedback loops*. In one of the presented control example it will be shown that the behaviour of a manufacturing plant can totally change and even safety constraints can be suddenly violated, if function blocks and applications are moved without the necessary care from one runtime and hardware platform to another. As reported in literature [FV04, CLA06, SZC+06, TD06a, VDVF07, DT08, YRVS09] the reason for this lies in the verbose description for the function block execution and the scheduling of function blocks inside function block networks. This leads to several runtime environments as FBRT [SZC+06], RTSJ-AXE [TZ05], RTSI-AXE [DBCT07], FORTE [Zoi09], Fuber [CLA06] and ISaGRAF [CB06a] complying with the standard, but using different semantics. Beside this, there exist an ANSI C based execution environment described in [CCB06] and the idea of an execution environment based on sequential or parallel hypothesis [VD07].
The known problem of function block execution is the firing of EC transitions if more than one clears. This problem can be solved by assigning priorities for each EC transition, because the engineer works with a graphical user interface and retrieving the evaluation order from the textual specification as mentioned in the standard can lead to a non-deterministic execution [TD06a].
The other problem of different scheduling possibilities of function blocks inside function block networks has to be examined by the control engineer himself, by knowing the scheduling of function blocks of the used runtime environment.
Possible runtime implementations reach from single-threaded resources with a concurrent, through multi-threaded resources with a parallel function block execution up to a true synchronous one. As outlined in [SZC+07] the single-threaded resource implementation could be done by *common function calls* as done at the FBRT and realising a kind of event stack, or by one or more *first-in-first-out queues* done by FORTE[1], Fuber and the one de-

[1]Only if the real time function blocks *RT_ ** are not used. Otherwise, a multi-threaded event-chain concept is used.

scribed in [CCB06]. Also the multi-threaded resource implementations can be subdivided into a *static* (RTSJ-AXE) or a *dynamic* (RTSI-AXE) assignment of function blocks to the threads. The synchronous scheduling approach is presented at [YRVS09] and based on the synchronous paradigm and implemented at Esterel. Additionally, there exist also a cyclical function block scan based implementation as ISaGRAF, which could be done with single or multi-threaded resources as well.

1.2 Proposed Solution Approach

Obviously, many different scheduling models have to be kept in mind if a truely portable application between several runtime environments should be developed. Even for an experienced control engineer this is not manageable in everyday's practice. Furthermore, not every different scheduling of function blocks influences the behaviour of the manufacturing plant and leads to malfunctions. Thus, formal models have to be created automatically or semi-automatically for the distributed control system and the plant to integrate the verification of the closed-loop system into the control engineering practices ([VH03]).

An appropriated way to model such closed-loop systems is the use of discrete-event systems as finite-automata, statecharts or Petri nets [HZ07]. Already in [ZD93] it is described how systems consisting of several components can be modelled with Petri nets . But, the major focus lies on the overall system behaviour, that is why a model of actuators and sensors is neglected. Another work of Zhou [Zho95] is focused on the transformation of Ladder Logic into Petri nets . An approach to cope with the state explosion is proposed in [ZV99] by reducing certain structures as single or simultaneous sequences of transitions and places in advance to the reachability analysis. To detect deadlocks during the resource allocation [ZF04] describes different strategies to find and avoid them by a liveness-enforcing supervisor.

The first approach for formal modelling of function blocks is presented in [VH99] and extended at the articles [VH00c, VH00a]. As formal model the *Net Condition/Event Systems* (NCES) are used and each state machine presented at the standard is represented by a separate NCE module. Thus, the model of a basic function block includes 6 different module instances describing the execution behaviour of the corresponding function block element. Thereby, the module instances describing an *event input*, an *event output* as well as an *EC state* are used as often as these elements exist. After connecting the modules the new composite module describes the execution behaviour of the basic function block. The used software package VEDA and the idea of an automatic verification framework is presented at [VH01c, VH01d].

Using the *Signal Interpreted Petri Nets* (SIPN) [WW00] presents an approach to model the event propagation through the function blocks by the token flow from a subnet to another. Each subnet is used as a black box with several input and output places. As common by SIPN the data flow is modelled separately to the Petri net and the algorithms of basic function blocks are modelled as functions to places.

A formal model based on the *synchronous dataflow language SIGNAL* is proposed by [SFL02]. Each function block is translated into a SIGNAL process describing the behaviour and receiving and emitting the signals as the corresponding function block. The assumption of this model is the parallel execution of all function blocks with no priority for them and the algorithms. To simplify the model a union event clock is defined for all event inputs. To ensure that no event is lost, the internal clock of each process has to be faster than this event clock. After expressing the specification in SIGNAL syntax the verification can be done with the SLIDEX tool. The presented example copes with fives function blocks, where the property of *an emergency stop if all pumps have a failure* is verified. If the property is not true, a trace of sequential input events is presented.

A more general approach with *Coloured Petri Nets* and focus on the operation mode of a function block is presented at [WSS03]. Therein, it is checked if all function blocks of an application get from the initial *Out of Service* mode into its *target mode.* Thus, only the initialisation procedure is checked.

In [SG04] a translation of function blocks into *timed automata* with the focus on their execution behaviour is proposed. For this reason algorithms and data of basic function blocks are not considered. Instead algorithms are modelled by their processing time. Each basic function block is modelled by a set of automata including an *Execution Control Chart* (ECC) automata, the *event input* automata as well as two *synchronisation* automata. Each function block network model is composed of the function block models scheduled by a scheduling automaton. Thereby, the scheduling automaton incorporates the non-preemptive and non-prioritized scheduling policy.

An approach with interacting *finite automata* is presented in [CLA06] for two different runtime environments. The one is Fuber with a sequential function block scheduling done by *first-in-first-out queues* and the other uses *common function calls* and may be FBRT. This approach results in two different formal models, but the impact to the plant is not shown if the scheduling is changed. Thus, this approach is extended at [CA08]. Therein, non-deterministic function blocks representing the different plant behaviours are described and used for simulation purposes. Due to the possibility of translating function blocks to finite automata, a formal closed-loop system is retrieved. The main drawback of this approach is, the formal plant model does not represent the plant behaviour, but rather the execution behaviour of function blocks describing the plant behaviour, which is not the same. Also, the algorithm translation of this approach is limited to assignments.

Based on the logic programming language SWI-Prolog [Wie09] an approach is presented by [DVH06], where a closed-loop function block application is analysed, which does not receive events from the environment through service interface function blocks. Each element of the function blocks is mapped to a prolog term with the same tag name and each algorithm is represented by a predicate. As every prolog implementation the production system is based on the resolvability of production rules and due to the backtracking possibilities all concurrent function block executions are calculated and inserted into a reachability graph. On this directed graph several properties as well as CTL formulas could be proven. At the presented example deadlocks are found for the special concurrent function block execution *common function calls.*

Extending the first mentioned approach with *Net Condition/Event Systems* from 1999 [PV07] presents ${}_{D}$TNCE structures modelling algorithms using unsigned integer-valued data by mapping the integer values to the token number. The main drawback of this straight forward approach is the state explosion. For example the presented ${}_{D}$TNCE structure to subtract 18 from a variable B covers $2 * 18 + 3$ states. The presented ${}_{D}$TNCE structure to multiply a variable A by 9 cover $(1 + 9) * A + 3$ states. Nonetheless the transfer of the data value from one NCE module to another one is not trivial. Furthermore, only a comparison between a variable and a fixed value could be done. In [PV08] the approach is extended to function block applications with the assumption of a sequential scheduling model.

In [Sue08], *Net Condition/Event Systems* are used to model the real time behaviour and a reconfiguration process of a function block network with the focus to the FORTE runtime implementation. This approach includes a model for the resource and the timing thread, but due to the exact modelled temporal behaviour, which is also done for the event propagation between function bocks, a trajectory for a control example with 0,9s duration gets at least 900000 states.

Concerning a formal definition of function blocks and applications [DV08] proposes an approach based on the set theory. Another approach limited to the ECC and including the interface is presented at [HW08] and used to determine *dead, transient, stable* and *semi-stable* EC states. Additionally these EC states can be subdivided into *conflicting, potential dead* and *pseudo semi-stable.* The used software is implemented in Java and searches through all function blocks of a provided system and prints out the state informations.

The main drawback of the most approaches is the absence of an appropriate plant model, because safety and production process specification are mainly provided by using plant and workpiece properties. Some other approaches neglect the modelling of algorithms, but the execution behaviour maybe influenced by executing algorithms as well. Nonetheless, the approaches with *Net Condition/Event Systems* seem to be the most promising approaches, because the hierarchical structure of the NCE modules can be identical to the one of the function blocks. Also once transformed NCE modules can be used over and over again as module instances. Although *Net Condition/Event Systems* is a concurrent formalism it is possible to model a sequential function block behaviour.

1.3 Structure of the thesis

To overcome the limitations of the presented approaches this thesis will present the idea of plant modelling by using a library of often used modules describing the causal and temporal behaviour of mechanical components at the end of Chapter 2. But as can be seen in Figure 1.1 the syntax and semantic definition for the used formal modelling language *discrete timed Net Condition/Event Systems* as well as their implementation in a workbench are presented before. In the following Chapter 3 several control implementation approaches, created during the thesis are presented and used to develop the transformation rules of function blocks, applications and resources in Chapter 4. In Chapter 5 the implementation of the transformation rules as an additional SWI-Prolog module of the workbench is described shortly. This is followed by modelling of the device and the plant as well as the interconnection to a closed-loop model. The results of the verification and for the different scheduling possibilities are described in Chapter 6. At the end the thesis and its results will be summarized.

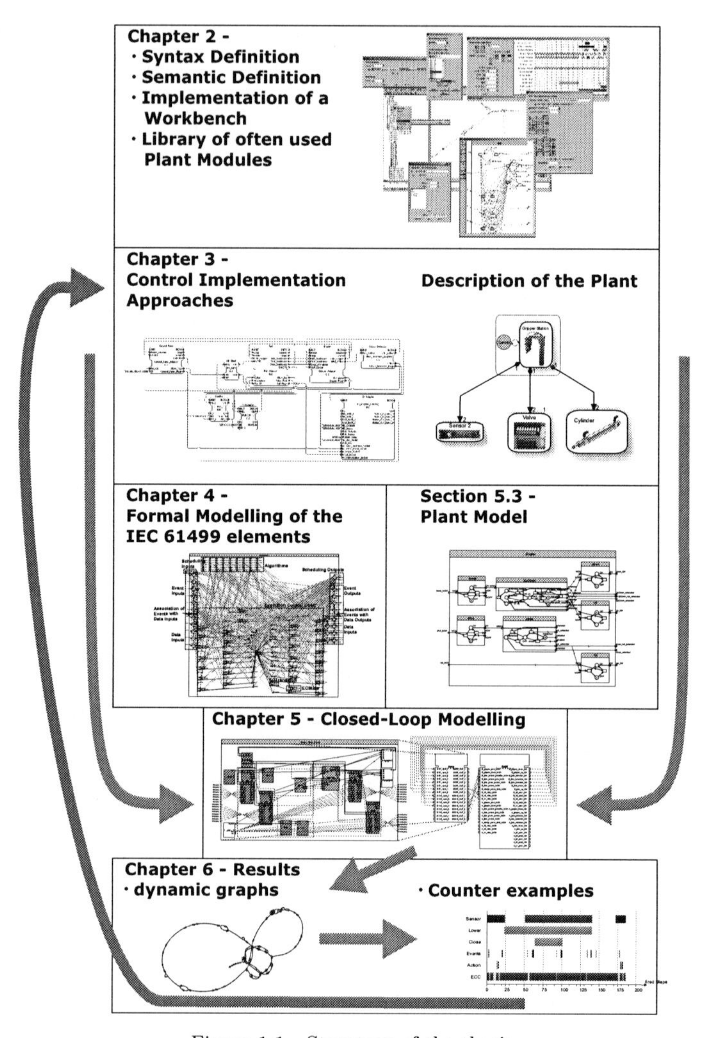

Figure 1.1 Structure of the thesis

Chapter 2

Formal Modelling Language

The chosen formal modelling language are the *discrete timed Net Condition/Event Systems* ($_D$TNCES), because they are modular and once developed modules can be used as instances over and over again. Thus, it will be possible to derive the hierarchical structure at the controller modelling from provided function blocks of an application and at plant modelling from the assembling of the used mechatronic components. Also the module interface with event in- and outputs at the head and condition in- and outputs at the body is quite similar to the one of function blocks [Kar09].
To model $_D$TNCE modules the TNCES-Editor was developed by the Chair for Automation Technology. With it also a lot of Petri net derivats can be modelled, but instead of the most Petri net editors[1] the TNCES-Editor is not able to perform any analysis, reachability calculation or simulation. Therefore the author implemented an expert system in SWI-Prolog [Wie09]. It will be described shortly after the presented syntax and semantic definitions of the formal modelling language. The main part of the expert system is a graphical user interface created with XPCE, the graphical toolkit included in SWI-Prolog. Beside this, several prolog modules to parse, create, analyse and view $_D$TNCE modules exist and are included in the *TNCES-Workbench*, but they can be used separately too.
Beside the working draft of the International Standard [ISO-IEC-15909-2] about a transfer format for high level Petri nets the author uses the existing TNCE markup language, to represent additional modelling elements as *modules* and *module networks* for structuring and reusability of once developed $_D$TNCE modules. The structuring element is the $_D$TNCE module with certain in- and outputs, providing information about the module state and occurring state changes. At the next higher level these module in- and outputs are connected through signal arcs transferring the condition and event signals from the source to the sink module. The limitation to discrete times should reduce the state explosion.

---

[1]

- SIPN (University of Kaiserslautern, http://www.eit.uni-kl.de/litz/ENGLISH/software/SIPNEditor.htm)
- ViEd+ViVe (University of Auckland, http://www.fb61499.com/valid.html)
- TimeNET (Technical University Ilmenau, http://www.tu-ilmenau.de/fakia/TimeNET.timenet.0.html)
- Petri-Netz-Editor (Vienna University of Technology,http://www.auto.tuwien.ac.at/~jlukasse/PED_Download.html)

2.1 Syntax Definition

According to [Thi02] and [Kar09] a *discrete timed Net Condition/Event Module* ($_D$TNCE module) is defined inductively by using a *discrete timed Net Condition/Event Structure* ($_D$TNCE structure) and extending it to a *base $_D$TNCE module* with condition and event in- and outputs as well as condition and event in- and output arcs. Using this *base $_D$TNCE module, composite $_D$TNCE modules* and *discrete timed Net Condition/Event Systems* (*$_D$TNCE systems*) can be composed by interconnecting the modules via event and condition interconnections.

Definition 2.1.1 *(Net, [Kar09])*
A net N is defined by the following tuple $N = (P, T, F)$ where:

- *P is a finite, non-empty set of places,*
- *T is a finite, non-empty set of transitions and $P \cap T = \emptyset$ and*
- *$F \subseteq ((P \times T) \cup (T \times P))$ is a set of flow arcs.*

□

Definition 2.1.2 *(NCE structure)*
Using the definition of a net, the Net Condition/Event Structures (NCE structure) can be defined as the following tuple

$$S = \{N, K, W_F, CN, W_{CN}, I, W_I, EN, em, sm\}$$

where:

- *$N = (P, T, F)$ is a net,*
- *$K : P \to \mathbb{N}^+$ defines a capacity for every place,*
- *$W_F : F \to \mathbb{N}^+$ defines an arc weight for every flow arc,*
- *$CN \subseteq (P \times T)$ is a finite set of condition arcs,*
- *$W_{CN} : CN \to \mathbb{N}^+$ defines an arc weight for each condition arc,*
- *$I \subseteq (P \times T)$ is a finite set of inhibitor arcs,*
- *$W_I : I \to \mathbb{N}^+$ defines an arc weight for each inhibitor arc,*
- *$EN \subseteq (T \times T)$ is a finite set of event arcs free of cycles, which means:*
  1. *$\nexists t \in T : (t, t) \in EN$ and*
  2. *$\nexists t_1, \ldots, t_i : (t_{l-1}, t_l) \in EN$ with $2 \leq l \leq i \wedge (t_i, t_1) \in EN$,*
- *$em : T \to \{\boxed{\wedge}, \boxed{\vee}\}$ defines an event mode for every transition $t \in T$ and*
- *$sm : T \to \{i, s\}$ defines a firing mode for every transition.*

□

The general firing mode of a transition is s (spontaneous) and is not noted in the graphical representation, and moreover it is only important for triggering transitions with no incoming event arc, as can be seen at the semantic definitions (Page 17). If a transition has no or only one incoming event arc, the event mode of the transition has no effect on the semantic, and due to this the event mode $\boxed{\wedge}$ is the general case and not noted in the graphical representation.
To ease the notations of further definitions, the author declares the following notations:

- ${}^{F}t = \{p \in P \mid (p,t) \in F\}$ is the set of all places at the pre-set and
- $t^{F} = \{p \in P \mid (t,p) \in F\}$ is the set of all places at the post-set of t.

The notations of the pre-and post-set of a place is done similarly.

Definition 2.1.3 *(Marking [Thi02])*
Let S be an NCE structure. Then m is a marking of S with the definition:

$$m : P \to \mathbb{N}_0, \qquad \text{and} \qquad \forall p \in P : m(p) \leq K(p).$$

□

Definition 2.1.4 *(${}_D$TNCE structure)*
A discrete timed Net Condition/Event Structure (${}_D$TNCE structure) is defined as the following tupel $S_T = (S, ZF)$ where:

- *S is an NCE structure and*
- *$ZF : ((P{\times}T){\cap}F) \to \{[a;b] \mid (a,b) \in (\mathbb{N}_0 \times \mathbb{N}_0^{\infty}) : a \leq b\}$ is non negative discrete time interval for every pre-arc (p,t) of a transition $t \in T$ with the following restrictions:*

$$ZF((p,t)) \to \begin{cases} [0;\infty] & \text{if } \exists t' \in T : (t',t) \in EN, \\ [ZF_R((p,t)); ZF_L((p,t))] & \text{otherwise.} \end{cases}$$

□

This means $ZF_R((p,t))$ is the retarding and $ZF_R((p,t))$ the limiting value of the time interval assigned to the flow arc $(p,t) \in (P \times T) \cap F$. Since not every transition without an incoming event arc has a time interval different from $[0;\infty]$ at its pre-arcs, it is used as general case and therefore not noted in the graphical representation.

Definition 2.1.5 *(Local time and state [Kar09])*
Let $S_T = (S, ZF)$ be a ${}_D$TNCE structure and m a marking of S. Then l is the local time to m with the definition:

$$l : P \to \mathbb{N}_0 \quad \text{and} \qquad \forall p \in P : l(p) = \begin{cases} 0 & \text{if } m(p) = 0, \\ 0 \ldots \infty & \text{otherwise.} \end{cases}$$

□

The local time is also named as age of the marking at place p [Han92a]. To describe the state of the ${}_D$TNCE structure S_T the following tupel $z = (m, l)$ will be used.

Definition 2.1.6 *(Maximal Age [Thi02])*
The definition of the maximal age of a marking of a place $p \in P$ of a $_D$TNCE structure S_T is:

$$lim(p) := \begin{cases} lim_R(p) & if\ lim_R(p) > lim_L(p), \\ lim_L(p) & otherwise, \end{cases}$$

where $lim_R(p)$ and $lim_L(p)$ are defined as follows:

$$\begin{aligned} lim_R(p) &:= max\left\{\{ZF_R((p,t)) \mid t \in p^F\} \cup \{0\}\right\}, \\ lim_L(p) &:= max\left\{\{ZF_L((p,t)) \mid t \in p^F \wedge ZF_L((p,t)) \neq \infty\} \cup \{-\infty\}\right\}. \end{aligned}$$

□

To get an base $_D$TNCE module, the $_D$TNCE structure S_T has to be extended by an in- and output set as well as an in- and output structure. The in- and output set defines the interface of each $_D$TNCE module as follows:

$$\Phi = (C^{in}, E^{in}, C^{out}, E^{out}).$$

Therein, C^{in} and C^{out} define a finite set of condition in- and outputs and E^{in} and E^{out} define a finite set of event in- and outputs.
Since the publication of [Thi02] several experiences using the described formal modelling language have been made by the working group the author belongs to and it seems to be useful to define the weight of the later defined condition and inhibitor chains at the input arcs of the base module (sink) instead of at the output arcs inside the source module. Thus, the in- and output structure is defined as follows:

Definition 2.1.7 *(In- and output structure)*
Let S_T be a $_D$TNCE structure and Φ an in- and output set. Then both can be connected by the following in- and output structure

$$\Psi(S_T, \Phi) = (CN^{in}, W_{CN^{in}}, I^{in}, W_{I^{in}}, EN^{in}, CN^{out}, EN^{out})$$

where:

- $CN^{in} \subseteq (C^{in} \times T)$ *is a set of condition input arcs,*
- $W_{CN^{in}} : CN^{in} \to \mathbb{N}^+$ *defines an arc weight for each condition input arc,*
- $I^{in} \subseteq (C^{in} \times T)$ *is a set of inhibitor input arcs,*
- $W_{I^{in}} : I^{in} \to \mathbb{N}^+$ *defines an arc weight for each inhibitor input arc,*
- $EN^{in} \subseteq (E^{in} \times T)$ *is a set of event input arcs, with the constraint* $\forall (p,t) \in F \mid \exists (e^{in}, t) \in EN^{in} : ZF((p,t)) = [0; \infty]$,
- $CN^{out} \subseteq (P \times C^{out})$ *is a set of condition output arcs, with the constraint* $\forall c^{out} \in C^{out} : \left|\{p \in P \mid (p, c^{out}) \in CN^{out}\}\right| \leq 1$ *and*
- $EN^{out} \subseteq (T \times E^{out})$ *is a set of event output arcs, with the constraint* $\forall e^{out} \in E^{out} : \left|\{t \in T \mid (t, e^{out}) \in EN^{out}\}\right| \leq 1$.

□

Definition 2.1.8 *(Base $_D$TNCE module)*
Every tuple $\mathcal{M}_B = (S_T, z_0, \Phi, \Psi(S_T, \Phi))$ is a base discrete timed Net Condition/Event Module, whereby z_0 defines the initial state of S_T. □

Using the definition of an base $_D$TNCE module one can inductively define composite $_D$TNCE modules, consisting of base $_D$TNCE modules or other composite $_D$TNCE modules and their interconnection by condition and event interconnections. Thus, it is possible to describe a hierarchy inside the $_D$TNCE modules.

Definition 2.1.9 *($_D$TNCE module)*
The discrete timed Net Condition/Event Modules are defined inductively as follows:

1. *Every tuple $\mathcal{M}_B = (S_T, z_0, \Phi, \Psi(S_T, \Phi))$, which is a base $_D$TNCE module is a $_D$TNCE module too.*

2. *If $\{\mathcal{M}_1, \mathcal{M}_2, \ldots, \mathcal{M}_k\}$ is a finite and non-empty set of $_D$TNCE modules, then is*

$$\mathcal{M}_C = (\{\mathcal{M}_1, \mathcal{M}_2, \ldots, \mathcal{M}_k\}, \Phi, CK, EK)$$

   a $_D$TNCE module too if

   - $\Phi = (C^{in}, E^{in}, C^{out}, E^{out})$ *is an in- and output set,*
   - $CK \subseteq \bigcup_{i\in\{1,\ldots,k\}} (C^{in} \times C_i^{in}) \cup \bigcup_{i,j\in\{1,\ldots,k\}} (C_i^{out} \times C_j^{in}) \cup \bigcup_{i\in\{1,\ldots,k\}} (C_i^{out} \times C^{out})$
     describes a condition interconnection within $\mathcal{M}_C$ with the constraint
     $$\forall c_s \in \left(C^{out} \cup \bigcup_{i\in\{1,\ldots,k\}} C_i^{in} \right) : \left|\{c_q \mid (c_q, c_s) \in CK\}\right| \leq 1 \text{ and}$$
   - $EK \subseteq \bigcup_{i\in\{1,\ldots,k\}} (E^{in} \times E_i^{in}) \cup \bigcup_{i,j\in\{1,\ldots,k\}} (E_i^{out} \times E_j^{in}) \cup \bigcup_{i\in\{1,\ldots,k\}} (E_i^{out} \times E^{out})$
     describes an event interconnection within $\mathcal{M}_C$ with the constraint
     $$\forall e_s \in \left(E^{out} \cup \bigcup_{i\in\{1,\ldots,k\}} E_i^{in} \right) : \left|\{e_q \mid (e_q, e_s) \in EK\}\right| \leq 1.$$

□

Each $\mathcal{M}_C$ is a *composite* $_D$TNCE module and the finite non-empty set $Sub(\mathcal{M}_C) = \{\mathcal{M}_1, \mathcal{M}_2, \ldots, \mathcal{M}_k\}$ describes the *submodules* of $\mathcal{M}_C$. The module set $Mod(\mathcal{M})$ of a $_D$TNCE module can be defined inductively as follows:

1. $\mathcal{M} \in Mod(\mathcal{M})$ and

2. $\mathcal{M}' \in Mod(\mathcal{M})$ if $\exists\, \mathcal{M}^* \in Mod(\mathcal{M}) \,:\, \mathcal{M}' \in Sub(\mathcal{M}^*)$

Through the inductive definition of a $_D$TNCE module it is ensured that it cannot incorporate itself ($\mathcal{M}_C \notin Sub(\mathcal{M}_C)$) and cyclic module instances are impossible. This means $\mathcal{M}'$ is not part of the module set $Mod(\mathcal{M}^*)$, if $\mathcal{M}^*$ is already a part of the module set $Mod(\mathcal{M}')$.

To ease the notations of the set of all transition and all places within a $_D$TNCE module $\mathcal{M}$, the following notations are use in the on going definitions:

- The set of all transitions $\overline{T}$ within $\mathcal{M}$ is: $\overline{T} := \bigcup_{\mathcal{M}_B \in Mod(\mathcal{M})} T_{\mathcal{M}_B}$ and
- the set of all places $\overline{P}$ within $\mathcal{M}$ is: $\overline{P} := \bigcup_{\mathcal{M}_B \in Mod(\mathcal{M})} P_{\mathcal{M}_B}$.

Definition 2.1.10 *(Event chain)*
Let $G_E(\mathcal{M}) = (V_E(\mathcal{M}), E_E(\mathcal{M}))$ be a directed graph of the $_D$TNCE module $\mathcal{M}$ with the vertices

$$V_E(\mathcal{M}) := \overline{T} \cup \bigcup_{\mathcal{M}_X \in Mod(\mathcal{M})} (E^{in}_{\mathcal{M}_X} \cup E^{out}_{\mathcal{M}_X})$$

and the edges

$$E_E(\mathcal{M}) := \bigcup_{\mathcal{M}_C \in Mod(\mathcal{M})} EK_{\mathcal{M}_C} \cup \bigcup_{\mathcal{M}_B \in Mod(\mathcal{M})} \left(EN^{in}_{\mathcal{M}_B} \cup EN_{\mathcal{M}_B} \cup EN^{out}_{\mathcal{M}_B}\right).$$

Then an event chain between two different elements $x, y \in V_E(\mathcal{M})$ (abbreviated $x \rightsquigarrow y$) exists, if there exists a path between x and y at the directed graph $G_E(\mathcal{M})$ [2].

Further it is required that event chains are free of cycles, which means:

1. *$\nexists t \in \overline{T} : t \rightsquigarrow t$ and*
2. *$\nexists t_1, \ldots, t_i \in \overline{T} : t_{l-1} \rightsquigarrow t_l$ with $2 \leq l \leq i \wedge (t_i \rightsquigarrow t_1)$,*

□

An event chain between two transitions exists, if there is a closed sequence of an event output arc, several event interconnections and an event input arc.

Definition 2.1.11 *(Condition- /Inhibitor chain)*
Let $G_C(\mathcal{M}) = (V_C(\mathcal{M}), E_C(\mathcal{M}))$ be a directed graph of the $_D$TNCE module $\mathcal{M}$ with all places and transition as well as all condition in- and outputs of all modules as vertices

$$V_C(\mathcal{M}) := \overline{T} \cup \overline{P} \cup \bigcup_{\mathcal{M}_X \in Mod(\mathcal{M})} (C^{in}_{\mathcal{M}_X} \cup C^{out}_{\mathcal{M}_X}),$$

and the connecting edges defined as follows:

$$E_C(\mathcal{M}) := \bigcup_{\mathcal{M}_C \in Mod(\mathcal{M})} CK_{\mathcal{M}_C} \cup \bigcup_{\mathcal{M}_B \in Mod(\mathcal{M})} \left(CN^{in}_{\mathcal{M}_B} \cup CN_{\mathcal{M}_B} \cup CN^{out}_{\mathcal{M}_B}\right).$$

[2] Path at the directed Graph $G_E(\mathcal{M})$:

$(x, y) \in E_E(\mathcal{M}) \vee \exists \{z_1, z_2, \ldots, z_i | \ (x, z_1) \wedge (z_i, y) \in E_E(\mathcal{M}) \wedge \ \forall 2 \leq i \leq n : (z_{i-1}, z_i) \in E_E(\mathcal{M})\} \subseteq V_E(\mathcal{M})$

Then a condition chain between two different elements $x, y \in V_C(\mathcal{M})$ (abbreviated $x \longrightarrow\!\!\bullet\; y$) exists, if there exists a path between them at the directed graph $G_C(\mathcal{M})$ of the $_D$TNCE module $\mathcal{M}$ [3]*.*

There exists an inhibitor chain between an element $x \in V_C(\mathcal{M})$ and a transition $t \in V_C(\mathcal{M})$ (abbreviated $x \multimap t$) if

$$(x,t) \in \bigcup_{\mathcal{M}_B \in Mod(\mathcal{M})} (I^{in}_{\mathcal{M}_B} \cup I_{\mathcal{M}_B}) \vee \exists z \in V_C(\mathcal{M}) \mid x \longrightarrow\!\!\bullet\; z \ : \ (z,t) \in \bigcup_{\mathcal{M}_B \in Mod(\mathcal{M})} (I^{in}_{\mathcal{M}_B}).$$

□

Thus, an inhibitor chain uses a condition output arc and several condition interconnections as well as an inhibitor input arc to connect a place and a transition.

As mentioned previously the weight of condition or inhibitor chain between a place and a transition is defined at the last segment of the connection.

Definition 2.1.12 *(Chain weights)*
Let $\mathcal{M}$ be a $_D$TNCE module with the in- and output set $\Phi = (C^{in}, E^{in}, C^{out}, E^{out})$ and $\mathcal{M}_B \in Mod(\mathcal{M})$ a base $_D$TNCE module, then the weight of a condition chain between $y \in V_C(\mathcal{M})$ and a transition $t \in T_{\mathcal{M}_B}$ ($y \longrightarrow\!\!\bullet\; t$) is:

$$W(y \longrightarrow\!\!\bullet\; t) := \begin{cases} max(W_{CN^{in}}((c^{in},t))) & if \ (c^{in},t) \in CN^{in}_{\mathcal{M}_B} \wedge (y \longrightarrow\!\!\bullet\; c^{in} \vee y = c^{in}) \\ W_{CN}((p,t)) & if \ (p,t) \in CN_{\mathcal{M}_B} \wedge y = p \end{cases}$$

and the weight of an inhibitor chain between $y \in V_C(\mathcal{M})$ and a transition $t \in T_{\mathcal{M}_B}$ ($y \multimap t$) is:

$$W(y \multimap t) := \begin{cases} min(W_{I^{in}}((c^{in},t))) & if \ (c^{in},t) \in I^{in}_{\mathcal{M}_B} \wedge (y \multimap c^{in} \vee y = c^{in}) \\ W_I((p,t)) & if \ (p,t) \in I_{\mathcal{M}_B} \wedge y = p. \end{cases}$$

□

This change to former definitions at [Thi02] was done due to several experiences using the described formal modelling language by the working group the authors belongs to and because it is better to decide at the receiving $_D$TNCE module how to use a provided information of another module. At the former definition the chain weight was defined at the condition output arc and thus only information about a lower or higher number of tokens at the connected place was provided to the environment of the $_D$TNCE module. If the marking of the place should be checked for several numbers of token, then all of them had to be modelled separately by a condition output arcs, the necessary condition interconnection and the input arcs.
If there exist more than one condition chain between the same elements, the maximum weight will be taken, according to the definition above. By more than one inhibitor chain

[3] Path at the directed Graph $G_C(\mathcal{M})$:

$(x,y) \in E_C(\mathcal{M}) \vee \exists \{z_1, z_2, \ldots, z_i | \ (x,z_1) \wedge (z_i,y) \in E_C(\mathcal{M}) \wedge \ \forall 2 \leq i \leq n : (z_{i-1}, z_i) \in E_C(\mathcal{M})\} \subseteq V_C(\mathcal{M})$

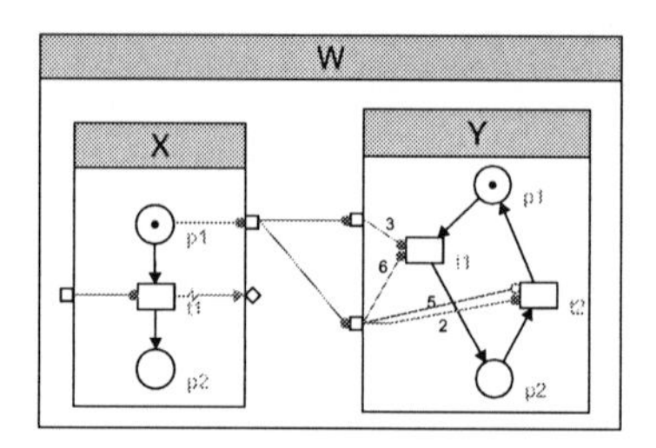

Figure 2.1 Weights of condition and inhibitor chains

the minimum weight will be taken. This leads to the following chain weights at the example shown in Figure 2.1:

$$
\begin{aligned}
p_1 \multimap t_1 \mid p_1 \in P_X, t_1 \in T_Y &: W(p_1 \multimap t_1) = 6,\\
p_1 \multimap t_2 \mid p_1 \in P_X, t_2 \in T_Y &: W(p_1 \multimap t_2) = 2 \textit{ and}\\
p_1 \multimap t_2 \mid p_1 \in P_X, t_2 \in T_Y &: W(p_1 \multimap t_2) = 5
\end{aligned}
$$

Definition 2.1.13 *(Marking, local time and state)*

- *A marking of a ${}_D$TNCE module $\mathcal{M}$ is defined as follows:*

$$\overline{m} : \overline{P} \to \mathbb{N}_0, \qquad \textit{and} \qquad \forall p \in \overline{P} : \overline{m}(p) \leq K(p).$$

- *A local time of a ${}_D$TNCE module $\mathcal{M}$ is defined as follows:*

$$\overline{l} : \overline{P} \to \mathbb{N}_0^\infty \quad \textit{and} \qquad \forall p \in \overline{P} : \overline{l}(p) = \begin{cases} 0 & \textit{if } \overline{m}(p) = 0, \\ 0 \ldots \infty & \textit{otherwise.} \end{cases}$$

- *The combination of the marking and local time of a ${}_D$TNCE module $\mathcal{M}$ to the tupel $\overline{z} = (\overline{m}, \overline{l})$ leads to the state of $\mathcal{M}$.*

□

Definition 2.1.14 *(Maximal age)*
The definition of the maximal age of a place $p \in \overline{P}$ within a ${}_D$TNCE module $\mathcal{M}$ is:

$$lim(p) := \begin{cases} lim_R(p) & \textit{if } lim_R(p) > lim_L(p), \\ lim_L(p) & \textit{otherwise,} \end{cases}$$

where $lim_R(p)$ and $lim_L(p)$ are defined as follows:

$$
\begin{aligned}
lim_R(p) &:= max\left\{\{ZF_R((p,t)) \mid t \in p^F\} \cup \{0\}\right\},\\
lim_L(p) &:= max\left\{\{ZF_L((p,t)) \mid t \in p^F \wedge ZF_L((p,t)) \neq \infty\} \cup \{-\infty\}\right\}.
\end{aligned}
$$

□

Definition 2.1.15 *(Event Source, Event Sink, Internal trigger transition)*
Let $\mathcal{M}$ be a $_D$TNCE module and $t \in \overline{T}$ a transition within, then

- *t is called event source if* $\exists x \in V_E(\mathcal{M}) \ : \ t \rightsquigarrow x$,
- *t is called event sink if* $\exists x \in V_E(\mathcal{M}) \ : \ x \rightsquigarrow t$,
- $\overline{^{EN}t} = \{x \in V_E(\mathcal{M}) \mid x \rightsquigarrow t\}$ *is the set of all event sources of t within* $\mathcal{M}$,
- $\overline{t^{EN}} = \{x \in V_E(\mathcal{M}) \mid t \rightsquigarrow x\}$ *is the set of all event sinks of t within* $\mathcal{M}$,
- $\overline{T_S} := \left\{t \in \overline{T} \mid \overline{^{EN}t} \neq \emptyset\right\}$ *is the set of event sinks of* $\mathcal{M}$,
- $\overline{T_{Ei}} := \left\{t \in \overline{T} \mid \overline{^{EN}t} = \emptyset \ : \ sm(t) = i\right\}$ *is the set of all internal instantaneous trigger transitions and*
- $\overline{T_{Es}} := \left\{t \in \overline{T} \mid \overline{^{EN}t} = \emptyset \ : \ sm(t) = s\right\}$ *is the set of all spontaneous trigger transitions.*

□

Every transition is an event source if it has an outgoing event chain to another transition or an event in- or output. It is an event sink if it has an incoming event chain. A transition with an incoming and an outgoing event chain is an event sink as well as an event source. Event chains serve as one-way synchronisation for firing a transition. An event sink is only allowed to be fired synchronously to its event source and have to if it is enabled. Depending on the event mode of an event sink $t \in \overline{T_S}$, it has to be fired if it has the event mode $em(t) = \boxed{\wedge}$ and if all event sources are active (see input state of an event input) or fired (transition). If the event sink has the event mode $em(t) = \boxed{\vee}$, it has to be fired if at least one event source is active or fired.

Definition 2.1.16 *(Incompletely controlled and ordinary transition)*
Let $\mathcal{M}$ be a $_D$TNCE module and $\Phi = (C^{in}, E^{in}, C^{out}, E^{out})$ the in- and outset, then

$$\overline{C^{in}}(\mathcal{M}) := \bigcup_{\mathcal{M}_X \in \{Mod(\mathcal{M}) - \{\mathcal{M}\}\}} C^{in}_{\mathcal{M}_x} \quad \text{and} \quad \overline{E^{in}}(\mathcal{M}) := \bigcup_{\mathcal{M}_X \in \{Mod(\mathcal{M}) - \{\mathcal{M}\}\}} E^{in}_{\mathcal{M}_x}$$

are the sets of enclosed condition and event inputs within $\mathcal{M}$[4].

The sets of all enclosed and connected condition and event inputs within the $_D$TNCE module $\mathcal{M}$ are defined as follows:

$$\overline{C^{in}_C}(\mathcal{M}) := \left\{c^{in} \in \overline{C^{in}}(\mathcal{M}) \mid \exists x \in \{C^{in} \cup \overline{P}\} : x \multimap c^{in}\right\},$$
$$\overline{E^{in}_C}(\mathcal{M}) := \left\{e^{in} \in \overline{E^{in}}(\mathcal{M}) \mid \exists x \in \{E^{in} \cup \overline{T}\} : x \rightsquigarrow e^{in}\right\}.$$

[4]These are the sets of all condition and event inputs within $\mathcal{M}$, without the inputs of $\mathcal{M}$.

A transition $t \in \overline{T}$ is incompletely controlled and an element of the set $\overline{T_U}$ if

1. $\exists c^{in} \in \left\{\overline{C^{in}}(\mathcal{M}) \setminus \overline{C^{in}_C}(\mathcal{M})\right\} : c^{in} \!\!-\!\!\bullet\, t \vee c^{in} \!\!-\!\!\circ\, t$ *or*

2. $em(t) = \boxed{\vee}$ *and*

$$\left(\forall e^{in} \in \overline{E^{in}}(\mathcal{M}) \mid e^{in} \rightsquigarrow t : e^{in} \notin \overline{E^{in}_C}(\mathcal{M})\right) \wedge \left(\forall t' \in \overline{T} \mid t' \rightsquigarrow t : t' \in \overline{T_U}\right) \text{ or}$$

3. $em(t) = \boxed{\wedge}$ *and*

$$\left(\exists e^{in} \in \overline{E^{in}}(\mathcal{M}) \mid e^{in} \rightsquigarrow t : e^{in} \notin \overline{E^{in}_C}(\mathcal{M})\right) \vee \left(\exists t' \in \overline{T} \mid t' \rightsquigarrow t : t' \in \overline{T_U}\right).$$

The set of ordinary transitions $\overline{T_G}$ is the complement of $\overline{T_U}$ in $\overline{T}$ ($\overline{T_G} = \overline{T} \setminus \overline{T_U}$). Inside a base module all transitions $t \in \overline{T}$ are ordinary transition and $\overline{T} = \overline{T_G}$. □

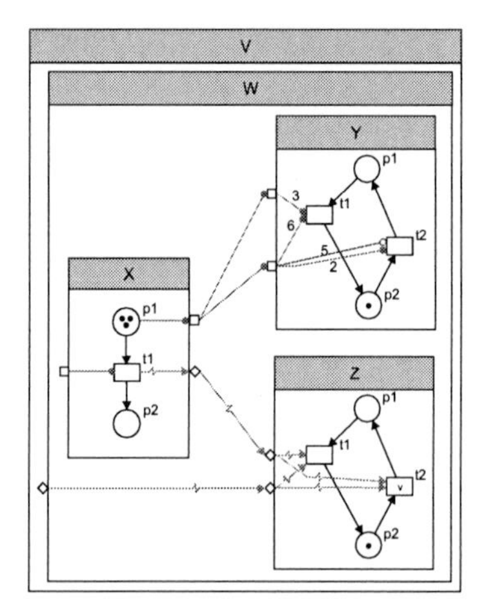

Figure 2.2 Incompletely controlled and ordinary transitions

This means at figure 2.2 transition $t_1 \in T_X$ is an ordinary transition of the $_D$TNCE module X, but an incompletely controlled transition at W and V. Furtheron, the transitions inside Y are ordinary transitions at every shown module, because their condition inputs are connected at W. The transition $t_1 \in T_Z$ is an ordinary transition inside Z, but due to the event mode $\boxed{\wedge}$ an incompletely controlled transition at W. Otherwise, the transition $t_2 \in T_Z$ is an ordinary transition at W and because of the unconnected event input at V an incompletely controlled transition at V.

Definition 2.1.17 *($_D$TNCE systems)*
A discrete timed Net Condition/Event System is defined as the following tuple $\mathcal{M}_S = (\underline{\mathcal{M}}, CK, EK)$ *where:*

- $\underline{\mathcal{M}} = \{\mathcal{M}_1, \mathcal{M}_2, \ldots, \mathcal{M}_k\}$ *is a finite and non-empty set of* $_D$*TNCE modules,*
- $CK \subseteq \bigcup_{i,j \in \{1,\ldots,k\}} (C_i^{out} \times C_j^{in})$
  describes the condition interconnection within $\mathcal{M}_S$ *with the constraint of* $\forall c_s \in \bigcup_{i \in \{1,\ldots,k\}} C_i^{in} : \big|\{c_q \mid (c_q, c_s) \in CK\}\big| \leq 1$ *and*
- $EK \subseteq \bigcup_{i,j \in \{1,\ldots,k\}} (E_i^{out} \times E_j^{in})$
  describes an event interconnection within $\mathcal{M}_S$ *with the constraint of* $\forall e_s \in \bigcup_{i \in \{1,\ldots,k\}} E_i^{in} : \big|\{e_q \mid (e_q, e_s) \in EK\}\big| \leq 1.$

□

If all modules of $\mathcal{M}_S$ are a base module, than $\mathcal{M}_S$ is called base system. The module set $Mod(\mathcal{M}_S)$ of a *discrete timed Net Condition/Event Systems* is defined as follows:

$$Mod(\mathcal{M}_S) := \bigcup_{i \in \{1,\ldots,k\}} Mod(M_i) \mid M_i \in \underline{\mathcal{M}}.$$

In contrast to the definitions of [Thi02] the $_D$TNCE systems do not contain any in- or outputs and describe the highest non-composable hierarchy level of a model. Naturally, that change is reflected in the model semantics at the following section, by replacing $\mathcal{M}$ by $\mathcal{M}_S$, except at the definition of the input state.

2.2 Semantic Definition

In the following section the previously defined syntax definition of $_D$TNCE modules shall be extended with the semantic definitions used for reachability calculation and simulation of $_D$TNCE modules and systems. To evaluate the enabling of a transition or step inside a $_D$TNCE module the interface to the environment has to be taken into account, which is done by the input state of the in- and output set.

Definition 2.2.1 *(Input state)*
For every $_D$*TNCE module* $\mathcal{M}$ *with* $\Phi = (C^{in}, E^{in}, C^{out}, E^{out})$ *is the event input state a mapping*

$$is_E : E^{in} \to \{True,\ False\}$$

and the condition input state a mapping

$$is_C : C^{in} \to \mathbb{N}_0^+,$$

assigning a value to each signal input. □

Definition 2.2.2 *(Extension of transition sets)*
Let $\mathcal{M}$ be a $_D$TNCE module with the in- and output set $\Phi = (C^{in}, E^{in}, C^{out}, E^{out})$ and $\xi \subseteq \overline{T_G}$ a non-empty set of transitions within $\mathcal{M}$.

$$\begin{aligned} Ext(\xi) :=& \Big\{ t \in \overline{T} \mid t \notin \xi \wedge em(t) = \boxed{\vee} : \\ & \quad (\exists t' \in \xi \mid t' \rightsquigarrow t) \vee (\exists e^{in} \in E^{in} \mid e^{in} \rightsquigarrow t : is_E(e^{in}) = True) \Big\} \cup \\ & \Big\{ t \in \overline{T} \mid t \notin \xi \wedge em(t) = \boxed{\wedge} : \\ & \quad (\{\forall t' \in \overline{T} \mid t' \rightsquigarrow t\} \subseteq \xi) \wedge (\forall e^{in} \in E^{in} \mid e^{in} \rightsquigarrow t : is_E(e^{in}) = True) \Big\} \end{aligned}$$

is the extension of the transition set ξ. □

This means a transition t with an event mode $\boxed{\vee}$ can only be part of the extension of the transition set ξ, if at least one event source $\overline{^{EN}t}$ is include in ξ or has the event input state $is_E(e^{in}) = True$. Otherwise, if the transition t has the event mode $\boxed{\wedge}$, it can only be part of the extension of the transition set ξ, if all connected event sources $\overline{^{EN}t}$ are part of ξ or have the event input state $is_E(e^{in}) = True$.

Definition 2.2.3 *(Step)*
The set of steps Ξ at the $_D$TNCE module $\mathcal{M}$ is defined inductively as follows:

1. *each trigger transition $t \in \{\overline{T_{Ei}} \cup \overline{T_{Es}}\}$ is a step*
2. *each event sink $t \in \overline{T_S}$ is a step if $t \in Ext(\emptyset)$*
3. *$(\xi \cup \{t\})$ is a step if ξ is a step ($\xi \in \Xi$) and $t \in Ext(\xi)$*

□

Everything else is not a step.
The notations ^Ft and t^F from the NCE structure can be used at $_D$TNCE modules. The set of places at the pre- or post-set is a subset of the places of the base $_D$TNCE module the transition $t \in \overline{T}$ belongs to. A derived notation can be used for the pre- and post-set of a transition set $\xi \subseteq \overline{T}$.

$$^F\xi = \{p \in \overline{P} \mid \exists t \in \xi \wedge \exists \mathcal{M}_{B/x} \in Mod(\mathcal{M}) : (p,t) \in F_X\}$$

$$\xi^F = \{p \in \overline{P} \mid \exists t \in \xi \wedge \exists \mathcal{M}_{B/x} \in Mod(\mathcal{M}) : (t,p) \in F_X\}$$

Definition 2.2.4 *(Marking and condition-enabled step)*
Let $\mathcal{M}$ be a $_D$TNCE module with the in- and output set $\Phi = (C^{in}, E^{in}, C^{out}, E^{out})$, the state $\overline{z}(\overline{m}, \overline{l})$ and the event input state is_E and the condition input state is_C. The transition $t \in \overline{T_G}$ is an ordinary transition and $\xi \subseteq \overline{T_G}$ is a non-empty set of ordinary transitions within $\mathcal{M}$.

- *t is marking-enabled at state $\overline{z}(\overline{m}, \overline{l})$ if*
  1. *$\forall p \in {^Ft} : \overline{m}(p) \geq W_F((p,t))$*
  2. *$\forall p \in t^F : K(p) \geq \overline{m}(p) + W_F((t,p)$*

- *t is condition-enabled at state $\overline{z}(\overline{m},\overline{l})$ and is_C if*

  1. $\forall p \in \overline{P} \mid p \multimap\!\!\bullet\, t \;:\; \overline{m}(p) \geq W(p \multimap\!\!\bullet\, t)$
  2. $\forall p \in \overline{P} \mid p \multimap t \;:\; \overline{m}(p) < W(p \multimap t)$
  3. $\forall c^{in} \in C_{in} \mid c^{in} \multimap\!\!\bullet\, t \;:\; is_C(c^{in}) \geq W(c^{in} \multimap\!\!\bullet\, t)$
  4. $\forall c^{in} \in C_{in} \mid c^{in} \multimap t \;:\; is_C(c^{in}) < W(c^{in} \multimap t)$

- *ξ is marking and condition-enabled at state $\overline{z}(\overline{m},\overline{l})$ and is_C if*

  1. *every $t \in \xi$ is marking and condition-enabled at state $\overline{z}(\overline{m},\overline{l})$ and is_C*
  2. $\forall p \in {}^{F}\xi \;:\overline{m}(p) \geq \sum_{\{(p,t)\in F_X \;\mid\; t\in\{\xi\cap T_X\}\}} W_{F_X}((p,t))$
  3. $\forall p \in \xi^{F} \;: K(p) \geq \overline{m}(p) + \sum_{\{(p,t)\in F_X \;\mid\; t\in\{\xi\cap T_X\}\}} W_{F_X}((t,p))$

The set Ξ_{mce} includes all ξ being marking and condition-enabled. □

Definition 2.2.5 *(Firing time)*
Let $\mathcal{M}$ be a ${}_D TNCE$ module with the in- and output set $\Phi = (C^{in}, E^{in}, C^{out}, E^{out})$, the state $\overline{z}(\overline{m},\overline{l})$ and the event input state is_E and the condition input state is_C. Further, $\xi \in \Xi_{mce}$ is a marking and condition-enabled step within $\mathcal{M}$.
The earliest firing time of a transition $t \in \overline{T}$ at $\overline{z}$ is:

$$fsz(t,\overline{z},is_E,is_C) := max\left\{\{ZF_R((p,t)) - \overline{l}(p) \mid p \in {}^{F}t \wedge \overline{m}(p) \neq 0\} \cup \{0\}\right\}.$$

and the latest firing time of a transition $t \in \overline{T}$ at $\overline{z}$ is:

$$ssz(t,\overline{z},is_E,is_C) := min\left\{\{ZF_L((p,t)) - \overline{l}(p) \mid p \in {}^{F}t \wedge \overline{m}(p) \neq 0\} \cup \{\infty\}\right\}.$$

The firing time τ of a marking and condition-enabled step ξ at $\overline{z}, is_E$ and is_C of $\mathcal{M}$ is depending on the included trigger transition or is 0.

$$\tau(\xi,\overline{z},is_E,is_E) := \begin{cases} fsz(t,\overline{z},is_E,is_E) & if\ t \in (T_{Ei} \cup T_{Es}) \cap \xi \\ 0 & otherwise \end{cases}$$

□

Definition 2.2.6 *(Enabled step)*
Let $\xi \in \Xi_{mce}$ be a marking and condition-enabled step of the ${}_D TNCE$ module $\mathcal{M}$ at state $\overline{z}(\overline{m},\overline{l})$ and is_C. Then ξ is called enabled if

1. $\nexists \xi' \in \Xi_{mce} : \tau(\xi',\overline{z},is_E,is_E) < \tau(\xi,\overline{z},is_E,is_E) \wedge \xi \subset \xi'$ *and*
2. $|\xi \cap T_{Ei}| = 1 \vee \forall \xi' \in \Xi_{mce} \mid \tau(\xi',\overline{z},is_E,is_E) = \tau(\xi,\overline{z},is_E,is_E) \;:\; |\xi' \cap T_{Ei}| = 0$.

□

An enabled step has to have the lowest firing time and has to be a maximal step at the given state and input state of $\mathcal{M}$. If there is a maximal instantaneous step with the lowest firing time marking and condition-enabled, this one gets enabled. Only, if there is no instantaneous step with the same firing time marking and condition-enabled, a maximal spontaneous step gets enabled. An enabled step is in the sense of Petri nets free of conflicts and contacts.

Definition 2.2.7 *(Successor state)*
Let $\mathcal{M}$ be a $_D$TNCE module with the in- and output set $\Phi = (C^{in}, E^{in}, C^{out}, E^{out})$, the state $\overline{z}(\overline{m}, \overline{l})$ and the event input state is_E and the condition input state is_C.
By firing an enabled step ξ at the time τ the successor state $\overline{z'}(\overline{m'}, \overline{l'})$ is reached and it is calculated as follows:

$$\begin{aligned}
\forall p \notin ({}^F\xi \cup \xi^F) \quad &: \quad l'(p) = \begin{cases} l(p) + \tau & if\ m(p) \neq 0 \wedge l(p) + \tau \leq lim(p) \\ lim(p) & if\ m(p) \neq 0 \wedge l(p) + \tau > lim(p) \end{cases} \\
\forall t \in \xi \quad &: \quad \forall p \in {}^F t : m'(p) = m(p) - W_F((p,t)) \wedge l'(p) = 0 \\
&\quad\ \ \forall p \in t^F : m'(p) = m(p) + W_F((t,p)) \wedge l'(p) = 0 \\
&\quad\ \ \forall p \notin \{{}^F t \cup t^F\} : m'(p) = m(p)
\end{aligned}$$

As abbreviation $\overline{z}[(\xi, \tau)\rangle\overline{z'}$ can be used. □

Definition 2.2.8 *(Firing sequence / Reachable states)*
Let $\mathcal{M}$ be a $_D$TNCE module. A state $\overline{z''}$ of $\mathcal{M}$ is called a follower state of $\overline{z_0}$ at the firing sequence $\overline{w} := (\xi_1, \tau_1), \ldots, (\xi_n, \tau_n) \in (\Xi \times \mathbb{N}_0^+)^*$ if

1. $|w| = 0 \wedge \overline{z_0} = \overline{z''}$ *or*
2. $\exists \overline{z'} \mid \overline{z_0}[(\xi_1, \tau_1), \ldots, (\xi_{n-1}, \tau_{n-1})\rangle\overline{z'} : \overline{z'}[(\xi_n, \tau_n)\rangle\overline{z''}$.

In this case $\overline{w}$ is called activated at $\overline{z_0}$ and $\overline{z_0}[w\rangle$ is written.
The set of all feasible firing sequences $\mathcal{W}_\mathcal{M}$ within $\mathcal{M}$ is:

$$\mathcal{W}_\mathcal{M} := \left\{\overline{w} \in (\Xi \times \mathbb{N}_0^+)^* \mid \overline{z_0}[w\rangle\right\}.$$

The set of reachable states $[\overline{z_0}\rangle_\mathcal{M}$ within $\mathcal{M}$ is:

$$[\overline{z_0}\rangle_\mathcal{M} := \{\overline{z} \mid \exists \overline{w} \in \mathcal{W}_\mathcal{M} \ : \ \overline{z_0}[w\rangle\overline{z}\}.$$

The set of executable steps $AS_\mathcal{M}$ within $\mathcal{M}$ is:

$$AS_\mathcal{M} := \left\{(\xi, \tau) \in (\Xi \times \mathbb{N}_0^+)^* \mid \exists \overline{z} \in [\overline{z_0}\rangle_\mathcal{M} \ : \ \overline{z}[(\xi, \tau)\rangle\right\}.$$

□

2.3 Markup Language

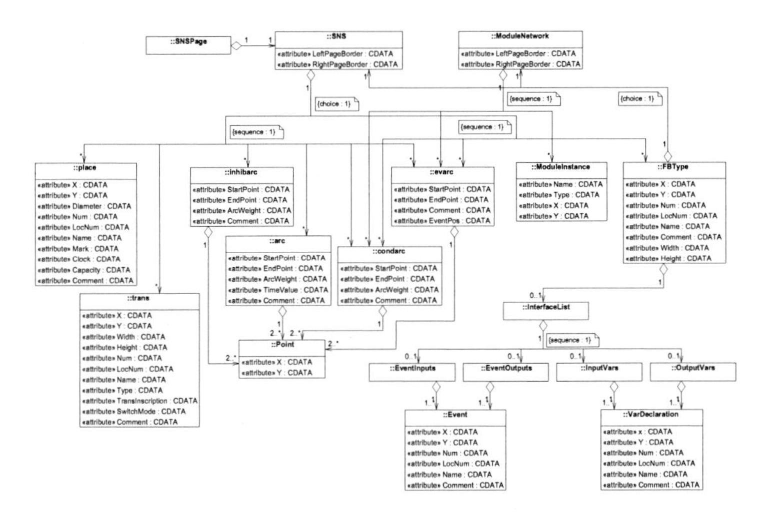

Figure 2.3 Class Diagram of the TNCES Markup Language

The described syntax at Section 2.1 of $_D$TNCE modules and systems are used to develop an XML based markup language to store, exchange and reuse once created modules. A file that meets the requirement of the markup language presented in Figure 2.3 is called *TNCEM or TNCES file.* According to the syntax definitions each $_D$TNCE module or system consists of the structuring objects base and/or composite *$_D$TNCE modules* coupled by signal interconnections via the event and condition in- and outputs. Inside the base $_D$TNCE modules form the elements *places, transitions and flow arcs* in conjunction with the signal arcs *event, condition and inhibitor* the underlying net. Thus, a $_D$TNCE module and $_D$TNCE system consists of mainly the same elements and for both filetypes *TNCEM or TNCES* the *TNCES markup language* consisting of the mentioned elements extended through the graphical elements *Point* and *SNS* as well as the *InterfaceList* can be used. The dependencies of all used elements are presented at an UML class diagram in Figure 2.3. The class diagram was developed according to the guidelines for Document Type Definitions (DTDs) of [CSF00] and created with Artisan Studio 7.0.19.

2.4 Implementation of a Workbench

During the thesis it was necessary to implement a new model checker, because the widely used *Signal Event System Analyzer* (SESA) does not fully support the modelling power of *discrete timed Net Condition/Event Systems.* Furthermore, the source code of SESA is written in a special programming language and compiled for the operating system Windows. Thus, it is not possible to use the Linux cluster at the Institute of Computer Science to handle any sizeable *discrete timed Net Condition/Event Systems.* Last but not least the support for SESA has been discontinued since the retirement of Professor *Dr. rer. nat. habil. H. Starke.*

The graphical user interface of the *TNCES-Workbench - V0.32* is implemented in SWI-Prolog [Wie09] and the graphical toolkit XPCE included in SWI-Prolog. This section shall provide a brief introduction of the workbench and the included modules. The graphical user interface of the workbench shown in Figure 2.4 is implemented in the module *nces-dialog*. This module incorporates several XPCE classes as any module implementing a graphical user interface. The class *nces_dialog* inside the mentioned module is derived from the class frame and incorporates a hierarchy browser at the left, a picture frame at the right and an editor frame at the bottom as well as a report dialog, a menu bar and a toolbar.
The *hierarchy browser* is used to show the modular structure of the parsed or due to some performed actions automatically created $_D$TNCE system. Beside, the frame of the type *graph_viewer* (derived from the class of *draw_canvas* and therefore from *picture*) shows the calculated dynamic graph, layouted as tree by SWI-Prolog itself or by using one of the provided layout algorithms of the *GraphViz* software [GKNV93]. For textual output, to load, to save and to edit the parameter file of each TNCEM or TNCES file, the *editor frame* below is used. At the bottom of the graphical user interface is the report dialog, which indicates the termination of each action by messages like "TNCES parsing finished...", "Calculation from state 0 terminated", "Dot Graph drawn" and so on.
During the parsing of a TNCEM or TNCES file the corresponding parameter file is checked for information about an external marking and external arc times as described in Section 6.1. This feature is useful if module instances are used over and over again, but the initial

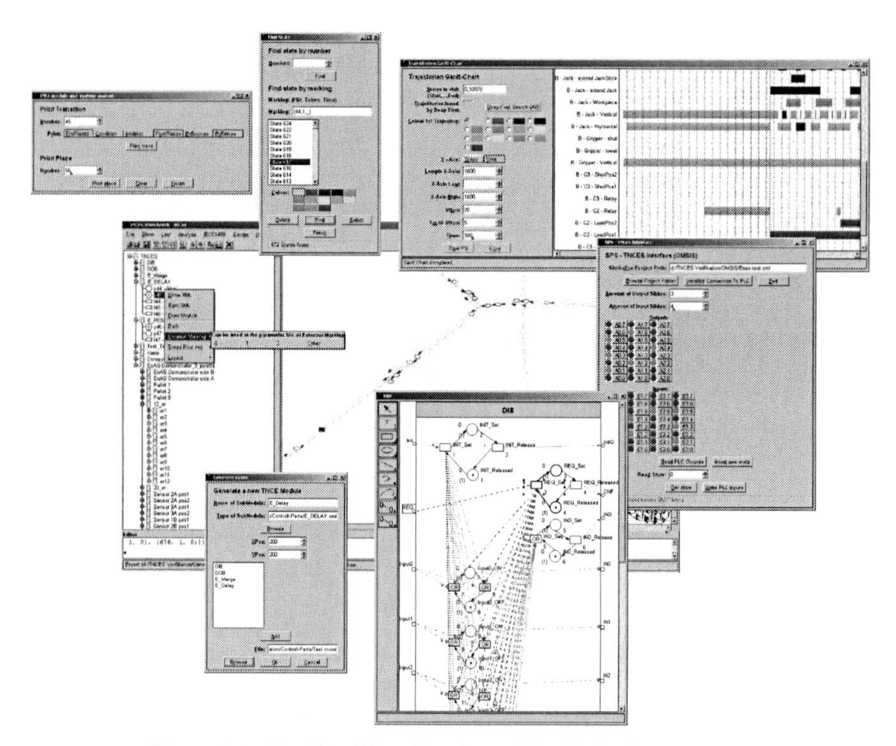

Figure 2.4 Graphical User Interface of the TNCES-Workbench

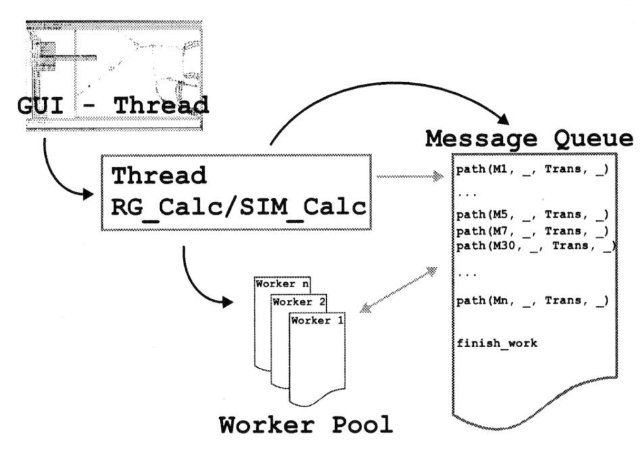

Figure 2.5 Worker-Pool-Model used during the reachability calculation

marking and the temporal behaviour should be different. After parsing a TNCEM or TNCES file a simulation can be done or the dynamic graph calculated. Because the dynamic graph is in most cases not only a sequence of states a worker-pool-model is chosen to boost the calculation of the state space for even huge models. As can be seen in Figure 2.5 each worker waits at the same message queue and takes the first predicate from it. The state space is calculated by a depth first implementation until a backtracking point is found and a certain depth limit is reached. Then a new predicate is inserted into the queue to start the state space calculation from this new state. After inserting the predicate the worker calculates all other states via the SWI-Prolog own backtracking routines and each time the depth limit is reached a new predicate is inserted into the message queue. If the message queue is empty and all workers are idle, the reachability calculation has finished and the predicate *finish_work/0* is inserted as often as workers are started into the message queue to kill all workers.

The difference between the reachability calculation and the simulation process as well as the resulting graph is described in detail at [Gut09]. The resulting graphs can be exported in dot syntax to a file, to be layouted by *GraphViz* [GKN06, GKNV93] [5] and imported back to the workbench or imported to the *ZGRViewer* [Pie05] [6]. The export and reimport of the graphs into the workbench can be achieved by the toolbar or the menu and the result is presented at the right picture frame. This opens the possibility to examine the graph further by applying state predicates, retrieving dead states, searching and colouring trajectories. Last but not least it is possible to use the clipboard of windows to easily exchange graphics between windows applications and to export the graphic into a postscript file.

Another simulation approach is described in [PGH10b] and implemented as a separate module of the workbench. It provides an online interface to Siemens PLCs and their remote I/Os via Profibus to read in the current output values and to write back new input values. The output values are mapped to the marking of the modules named *DO-X*. Afterwards, a reachability calculation is done and several states can be chosen to get new input values from the marking of the modules named *DI-X*. Doing this closed-loop simulation of the

[5] Webpage of the Graph Visualization Software http://www.graphviz.org/

[6] Webpage of the ZGRViewer, a GraphViz/DOT Viewer http://zvtm.sourceforge.net/zgrviewer.html

actual control code with a formal plant model can be done, to test if the controller reacts in any case appropriated to the specification.
The transformation rules presented in Chapter 4 are implemented in a module too, and therefore function blocks can be automatically transformed to NCE modules.
The distribution of the TNCES-Workbench is done via the *SourceForge project 341432* named *TNCES-Workbench* as ready to use stand alone windows application with all used *Dynamic Link Libraries* of the latest SWI-Prolog version and as source package to be used with a 32 or 64 bit version of SWI-Prolog either at a unix or windows platform. The *GraphViz* package is included in the corresponding folder and can be replaced by a newer version if necessary [7]. Other applications as the *ZGRViewer*, a *saxon8* XML parser used for XSLT transformation and the used XSLT scripts are included too.
A deeper description of the implementation and the features of the TNCES-Workbench can be found at the corresponding technical report or at the SourceForge project website.

2.5 Library of often used Plant Modules

The dominating part of each manufacturing system is the plant. The chosen controller, either hierarchical, central or distributed, is only a means to realize the manufacturing process and to fulfil necessary safety constraints. Thus, it is a doubtful activity to design a control system without a minimum knowledge of the physical plant behaviour and the desired production process. Nevertheless, it should be mentioned that designing a model of the plant used for simulation or formal verification is a time consuming task and therefore a cost factor that must not be underestimated. Designing a model from scratch is therefore not appropriate, and a methodology of engineering models in a systematic way rather than designing them from scratch is required. Therefore, a modular and compositional approach is helpful although it is not the ultimate solution of all problems. But having a look at the everyday engineering workflow, this closed loop is the best way to identify implementation errors in advance and therefore the plant will be brought into service faster.
Having a look to manufacturing systems, each of them consists of mainly the same mechatronic components as conveyors, cylinders, heaters, tanks, storage, etc. as well as of digital and analogue sensors and actuators. The sum of all components describes, the physical equipment of the plant, which changes the properties of the workpieces during the manufacturing process. Every component should be described by a module encapsulating the discrete and uncontrolled behaviour of the representing mechatronic component. This means it has to describe each behaviour which is possible at each physically possible state under each order of arbitrary assignment of the actuator states, as described in [HKL99]. During the systematic modelling of the plant components using prior developed module instances describing the basic components and stored in a library, a signal coupling by event and condition interconnections as described in [VH05b] has to be done. Doing this composition of large models from smaller ones is obvious to any engineer who has ever modelled a system in block-diagram-oriented way. In the following some basic plant modules are presented and used at Section 5.3 to model the whole plant.

[7]Download from `http://www.graphviz.org/` and for testing purpose the fact `graphviz_bin_dir('Graphviz/bin').` at the parameter file have to be updated

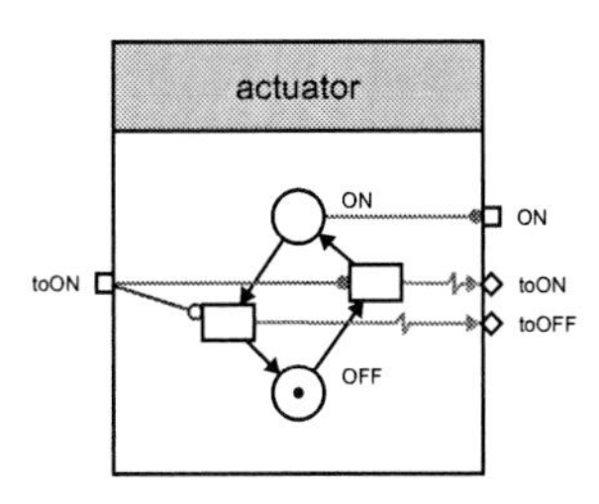

Figure 2.6 NCE module of an actuator

Mechanical components not instantly switched on

The elementary mechanical components not instantly switched on are valves, suckers, relays and drives with one rotating direction. All of them can be switched on and off to influence other actuators as cylinders or conveyors as well as several workpieces. The formal model in Figure 2.6 consists of a place invariant with the places *ON* and *OFF*. The condition input *toON* enables the transition connecting them. The condition output *ON* can be used to connect modules of other mechanical components, which are not instantly influenced by the actuator state changes as for example cylinders or drives. The event outputs *toON* and *toOFF* shall be used to model interconnections to mechanical components, which are instantly influenced by direct mechanical interaction as for example a conveyor by an drive or a workpiece by a sucker.

Mechanical components instantly switched on

The formal model of mechanical components instantly switched on consists of a place invariant as well, but the state transitions are enforced by input events. The NCE module in Figure 2.7a can be used to describe the behaviour of sensors or conveyors with a mechanical connection to the drive and for the local workpiece behaviour. The right NCE module of the same figure may describe the behaviour of a sensor, which will only be switched on if two constraints are fulfilled. If either the first constraint (⟶ *toON_1*) or the second constraint (⟶ *toON_2*) is fulfilled and the other one gets fulfilled too (⇝ *toON_2*, ⇝ *toON_1*), then the left or right transition will be enabled and the token flows at the same step to place *ON*. Due to the event inputs *toOFF_1* and *toOFF_2* the token returns to the place *OFF*, if one of the constraints is no longer fulfilled.

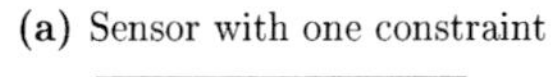
(a) Sensor with one constraint

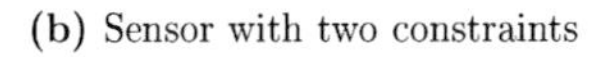
(b) Sensor with two constraints

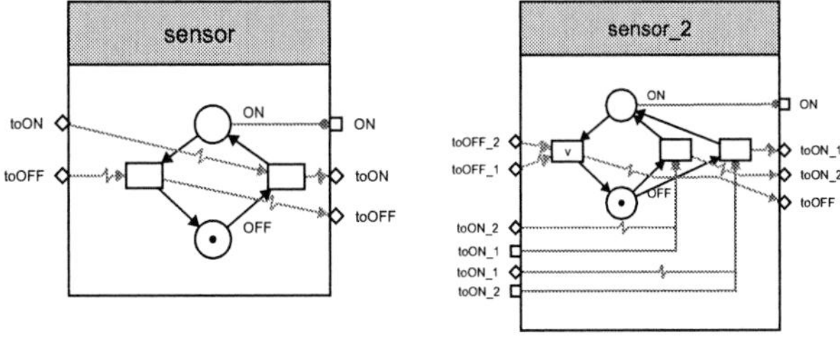

Figure 2.7 NCE modules of a sensor

Mechanical components with two end positions

Two members of this group of mechanical components are tanks and cylinders. A tank maybe empty, full or something between, and a cylinder is retracted, extended or at a position between both. If more discrete position are needed, additional places and transitions have to be inserted into the module. The cylinders used in the thesis are extended or retracted by compressed air. The opening of the corresponding valve enables the flow of air

(a) Without discrete time

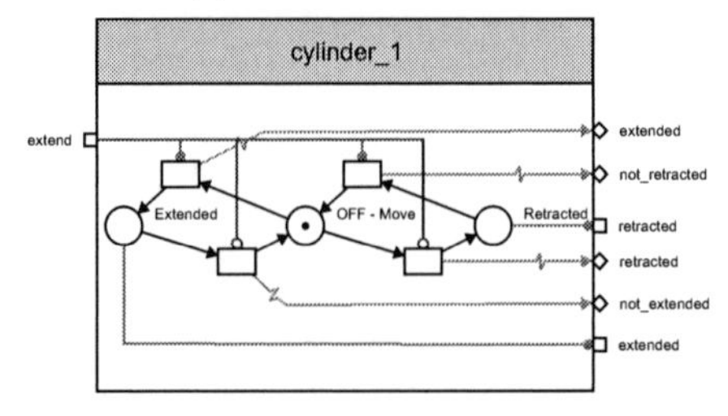

(b) With discrete time to leave the end positions

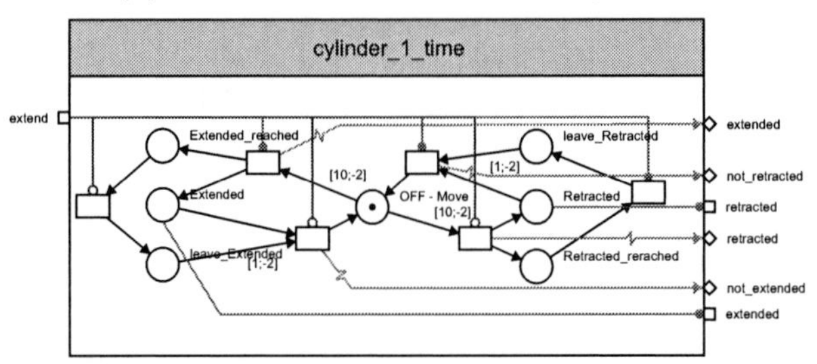

Figure 2.8 NCE module of a single-acting cylinder

into the cylinder and closing the valve, the air flows out again. Thus, the cylinder does not instantly retract or extend, if the valve is opened or closed because the change of pressure is time-dependent. This time depends on the system pressure, the length and diameter as well as the construction of the tube and cylinder (i.e. the number of tube connectors, corners, T and y-fittings). Consequently, the cylinder model can only be connected by condition arcs to other modules. In Figure 2.8a, this condition signal enables the state transitions between *Retracted*, *Move* and *Extended*.

Discrete times can also be inserted into the model, so that it is extended to a $_D$TNCE module. By considering a time to leave the end positions, the module shown in Figure 2.8b is extended by four places **_ reached* and *leave_ ** as well as two connecting transitions, whereas * stands for *Retracted* and *Extended*, respectively. The places * and **_ reached* get a token at the same step. If the condition signal changes, the transition between the place * and *Move* will only be enabled, if there is a token at the place *leave_ **, which gets there from the place **_ reached*. The condition enabling of the transition between the places * and *Move* as well as **_ reached* and *leave_ ** is the same. If now a discrete time is noted at the post-arc of the place *leave_ **, the token will also stay at * for the modelled time even though the transition is condition-enabled.

In Figure 2.9a and 2.9b the models of double-acting cylinders are presented. In contrast to the single-acting cylinder, there are two chambers, which have to be filled and released with compressed air to move the cylinder. In case that both connected valves are opened, both chambers are filled and the cylinder will remain in its position until one valve is closed. It is also possible to introduce discrete times into the formal model as previously described.

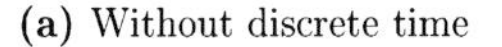

(a) Without discrete time

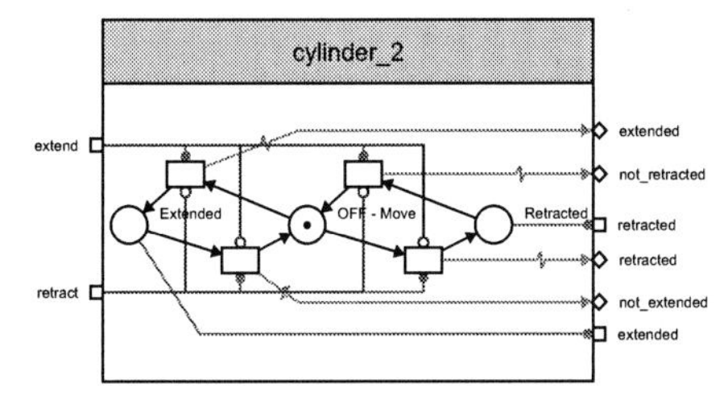

(b) With discrete time to leave the end positions

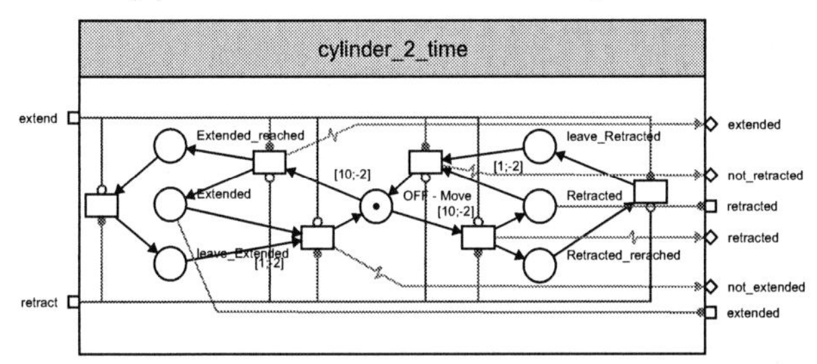

Figure 2.9 NCE module of a double-acting cylinder

2.6 Summary

This chapter starts with the syntax and semantic definitions of the later used formal modelling language $_D$TNCE systems. Thereby, the definition of the used local times is changed to a discrete one to reduce the resulting state space. Furthermore, the arc weights of condition and inhibitor connections are now defined by the input arcs and not by the output arcs as done in former publications.

Following, the used *TNCES Markup Language* used to exchange $_D$TNCE modules and systems between the existing tools at the Chair for Automation Technology is described. Following, the implementation of the syntax and semantic definitions in the expert system SWI-Prolog is described shortly. This reimplementation was necessary for the ongoing work, because the existing TNCES-Editor is not able to analyse the created $_D$TNCE modules and systems and the used modelchecker SESA does not fully support all modelling elements. Furthermore, the support and further development of SESA is cancelled and it does not support a parallel computation. Thus, the TNCES-Workbench was created and used to achieve the results presented in Chapter 6. Also the defined transformation rules of Chapter 4 are included into the workbench and thus it can be used in combination with the TNCES-Editor for the closed-loop modelling in Chapter 5.

Concluding this chapter a library of base $_D$TNCE modules is presented to describe the behaviour of often used plant components. Thus, the model of the plant has not to be developed from scratch, but by glueing modules together. Despite this, the model of the workpiece behaviour has to be developed every time from scratch, because it changes from plant to plant.

Chapter 3

Control Implementation Approaches

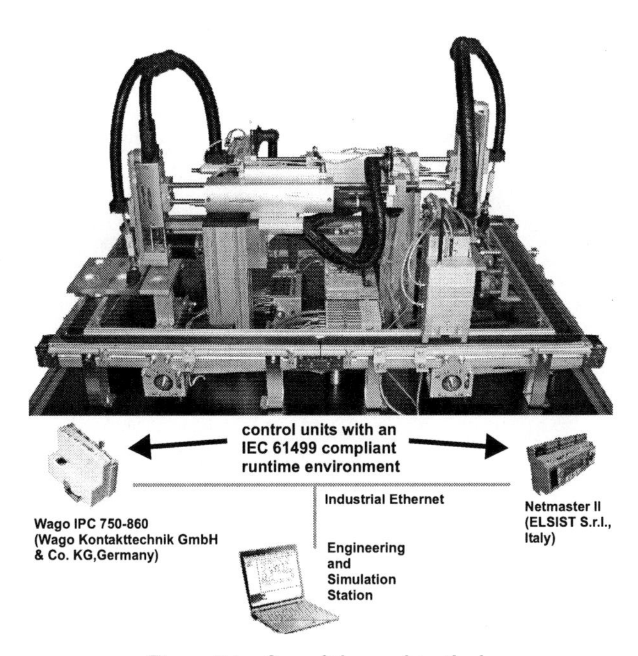

Figure 3.1 One of the used testbeds

This chapter shall provide an overview about possible implementations of distributed control systems developed and realised at the different testbeds of the Chair for Automation Technology and based on the approaches of [VHH06]. Since some of the testbeds were primarily used to introduce the programming of *programmable logic controllers* (PLC) to students, the first implementations were done as Central Controllers. Thereby, one monolithic function block realises the whole control application. From this first straightforward approach, several others are derived and presented in the following using the testbed shown in Figure 3.1.

As described in [PGH10b] the testbed consists of two identical plant parts rotated by 180° to each other. Both parts have a mounted *Jack, Slide* and *Gripper Station* to unload, load,

close and open tins. These tins are transported by the conveyor circuit on pallets between the processing stations. To position the tins correctly in front of the stations, each conveyor is equipped with two light barriers.
The *Jack Station* puts workpieces from the *Slide Station* to the tins on the pallet or vice versa. Furthermore, it can open a tin and deposit its lid onto the pallet or back onto the tin again. The main function of the *Gripper Station* is to close the lid of a tin, which was previously loaded by the *Jack Station.* Furthermore, a tin can be lifted and deposited onto another pallet.

3.1 Short Introduction to IEC 61499

[IEC-61499-1] defines several software models to implement distributed control systems. As presented in Figure 3.3, which is impressed by figure 2-2 of [Zoi09], these are from top to down the *system configuration*, the *device model*, the *resource model* and the four function block types *simple, basic, composite* and *service interface.* Despite this, an *application model* is defined to realise a development of control applications independently from the used control devices. After the development of an application, all parts of it are mapped to several resources and therefore to the control devices.

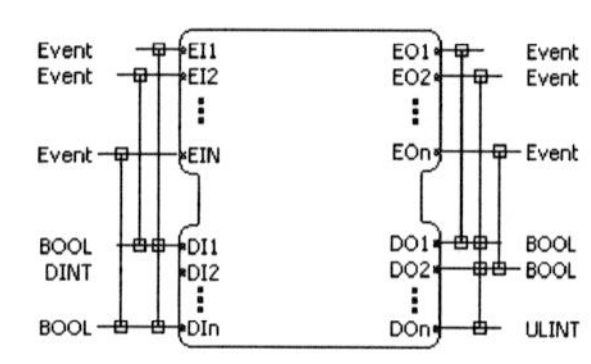

Figure 3.2 Function Block Interface

Starting with the basic elements the software model defined by the [IEC-61499-1] to implement distributed system should be described shortly. An extended description can be found at the standard itself or at the books [Lew01, Vya07, Zoi09]. As presented in Figure 3.2, the common part of all function blocks is the interface consisting of event and data in- and outputs to exchange information with their environment. If an event occurs at the inputs, the associated data inputs are sampled, and the function block is scheduled for execution. During the execution, the data inputs have to be consistent and are processed to produce none or more output events and new data output values. The new output values are only published, if an associated output event is published. As shown Figure 3.3d, the data processing differs among all function block types and is described next.

Simple function blocks:
Annex D of the [IEC-61499-1] defines the *simple function blocks* enabling the control engineer to use [IEC-61131-3] function blocks in an event-driven manner. During the necessary transformation, several algorithms are created, and for each algorithm an in- and output event is inserted to trigger the execution of the algorithms. Thus, a simple function block has only a set of algorithms to process the data inputs as presented in Figure 3.3d.

Basic function blocks:
In section 5.2 the standard defines the *basic function blocks* utilising an *Execution Control Chart* (ECC) to control the execution of its incorporating algorithms. As shown at the 2nd FB type of Figure 3.3d, the ECC consists of one or more *EC states* linked by *EC transitions.* Furthermore, each EC state has none or more associated *EC actions.* These

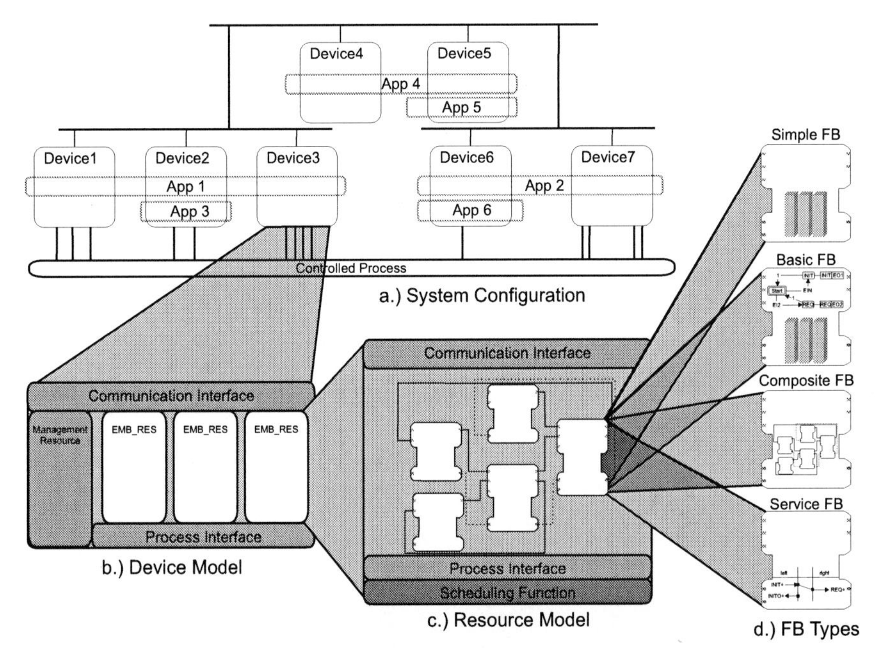

Figure 3.3 Elements of the IEC 61499 Software Model

EC actions consist of algorithms and output events, and they are executed once the EC state is activated. Each EC transition has a condition incorporating input events and data as well as internal data values. This condition is evaluated if the execution of the previous EC state is finished, and if an input event occurs and the previous EC state is active.

Composite function block:
Composite function blocks incorporate a set of component function blocks connected by event and data connections. Each input event and data is passed on to the connected component function blocks. Each output event or data of a component function block may be passed to another component function block or to the event and data outputs of the composite function block.

Service Interface function blocks:
Service Interface function blocks provide access to services provided by the underlying operation system of the control device. Some typical services are the read and write to the process interface (physical inputs and outputs of the device) and to the communication interface. The desired interaction of the control application with provided service is documented by service primitives in time sequence diagrams according to the [ISO-IEC-10731].

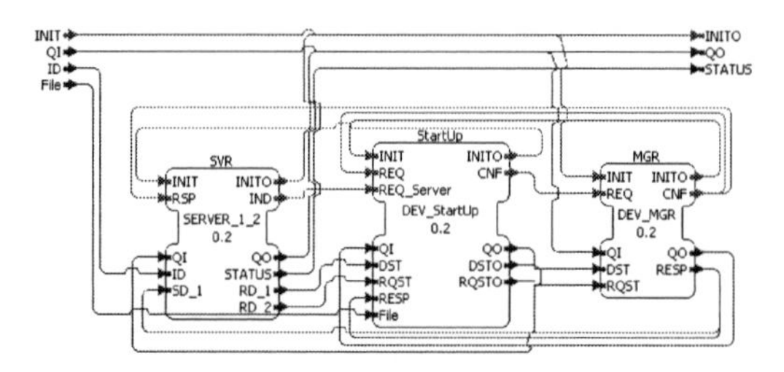

(a) Function Block Network of the new DM_KRNL function block

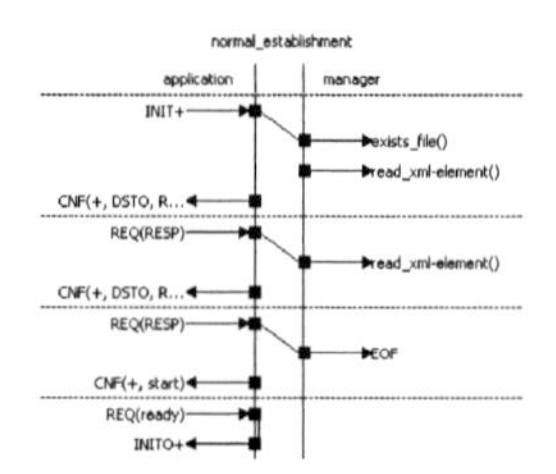

(b) Service Sequence of the normal initialisation process

Figure 3.4 DM_KRNL function block for an instant device start-up after power down

Applications:

Control applications can be developed using instances of once created function block types . Inside a control application, a subapplication may exist to group several function blocks together. Afterwards, each element of the control application is mapped to a resource and therefore to a device of the system. Thereby, the data exchange through the system communication network has to be established. As shown in Figure 3.3a, application 1 is mapped to the devices 1, 2 and 3, and application 6 runs only at device 6.

Devices:

Each control device of a system configuration incorporates at least the *communication and process interface* as well as a *management resource* as presented in Figure 3.3b. The communication interface provides communication services to the application parts mapped to this device, and the process interface provides services to read and write to the actuator and sensor of the device to control the process. The *management resource* enables the device to receive and encode management commands to *create, start, stop, delete* and *configure* other resources and function block instances as well as their event and data connections. Thus, each device provides a basic reconfiguration service for the engineering tools to manage them [Zoi09]. According to the *IEC 61499 Compliance Profile for Feasibility Demonstration,* [1] the management resource is named *RMT_RES* and includes the function blocks *E_Restart*, *E_SR* and *DM_KRNL*. Thereby, the function block *DM_KRNL* realises the management functions, but it has to be extended to the one presented in Figure 3.4 to realise an *instant start-up* of the device after power down, which is necessary for every industrial used device. The interface is extended by the data input *File* providing the path to a file located at the device and storing all management commands sent to the device. During the initialisation process at the next start-up of the device, command by command is read from this file. This procedure ensures that the device will run with the same configuration as the one before the power down occurs. Nonetheless, the device is manageable from any [IEC-61499-2] compliant engineering environment, and each additional management command is added to the configuration file as well.

[1] http://www.holobloc.com/doc/ita/index.htm

3.2 Central Controller Approach

The function block presented in Figure 3.5 is the *Central Controller* of the left plant part of the described testbed. The event output *Timer_Start* is used to start an external timer function block of the type *E_Delay* and the expiration of the timer is recieved through the event input *Timer_Fin*. The other event inputs *INIT* and *REQ* are connected directly to the service interface function blocks (SIFBs) reading the inputs of the device, and the output events *INITO* and *CNF* are connected to the SIFBs accessing the outputs of the device. Thus, this function block controls all 3 conveyors, the Jack, the Seldge and the Gripper Station. Therefore, the Execution Control Chart (ECC) shown right to the interface in Figure 3.5 gets huge and hard to maintain. Due to the requirements defined at the [IEC-61499-2] this function block is portable between several engineering environments, but it is not reusable if the production scenario changes. For example, if the first tin of the pallet is loaded and the second is unloaded at the left plant part, than the reverse operations have to be performed at the other part to get a circular production process. To realise this change of the production process the source and accordingly the sink of four EC transitions have to be updated to create the Central Controller of the other plant part. These EC transitions lie at the bottom of the green and red rectangle overlaying the ECC in Figure 3.5. The green (left) rectangle incorporates all EC states and EC transitions necessary to load a workpiece into the tin and to place the lid onto the tin again. Inside the red (right) rectangle all EC states and EC transitions are included to open and unload a tin. To mange the conveyors and to transport the pallets to the Jack and the Gripper Station as well as to the right plant part the EC states and EC transitions inside the brown polygon are used. During the up and down movement of the Gripper Station and the closing of a tin the EC states inside the lower blue rectangle are activated. Every stable EC state is connected by an EC transition with the condition *STOP* to the EC state *STOP*, to interrupt the actual production process. Thereafter, the function block has to be reset by receiving the *RESET* event and to reach the initial EC state again.
Using this information several actions can be identified for every used mechanical component and split to separate function blocks as described in [VHH06] for another plant. This leads to the approach of Master-Task-Controller presented at the following section.

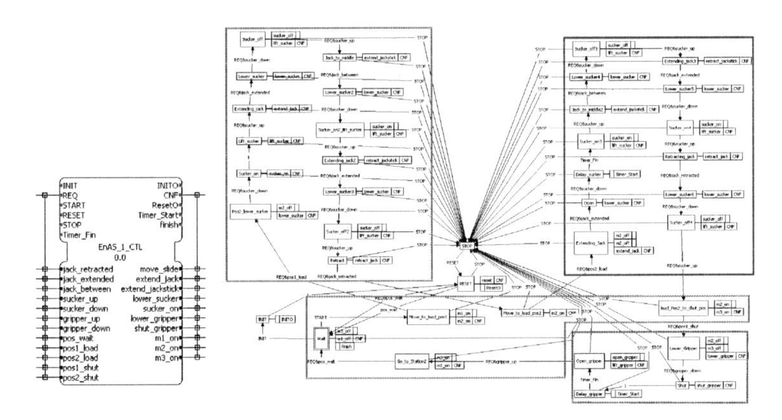

Figure 3.5 Central Controller of the left plant part

3.3 Master-Task-Controller Approach

As described in [VHH06, HGVH07], the once for every mechanical component developed Task-Controller is the reusable part of the control application and can be stored in a library of function blocks. If now a mechanical component is removed from or is additionally mounted to the plant, the function block of the Task-Controller has to be deleted from or inserted into the control application. Although, if the components are rearranged the function block has only to be reconnected.

In Figure 3.6 the Task-Controller of the Jack Station is shown. Again the green (left) rectangle incorporates all EC states getting sequentially activated during the loading of a workpiece into a tin. This elementary action of the Jack Station is triggered by the input input *load*, if the EC state *Wait* is active. Receiving the input event *unload* during the EC state *Wait* is active, the action to open and unload a tin runs through. Once again, every stable EC state is connected by an EC transition with the condition *Stop* to the EC state *Stop*, to stop the actual action.

The only drawback of this approach is the creation of the Master-Controller for every production scenario. In fact the changes are easier to realise than at the former presented approach, because only the actions of the Task-Controller have to be coordinated, but every new or changed function block has to be uploaded to the control device, instantiated and connected. But, not every runtime environment supports this dynamic reconfiguration as reported in [OWRSB05, VHH06]. Thus, the control application has to be killed and relaunched again. This leads to the approach of Parametrized Master-Task-Controller, where a set of mechanical components gets its own Master- and several Task-Controllers. The Master-Controller incorporates every possible production scenario and coordinates it with the other Master-Controllers of the plant.

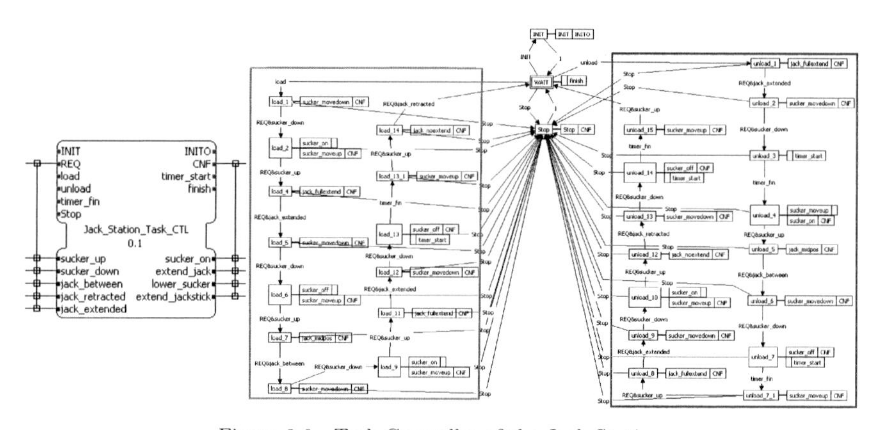

Figure 3.6 Task-Controller of the Jack Station

3.4 Parametrized Master-Task-Controller Approach

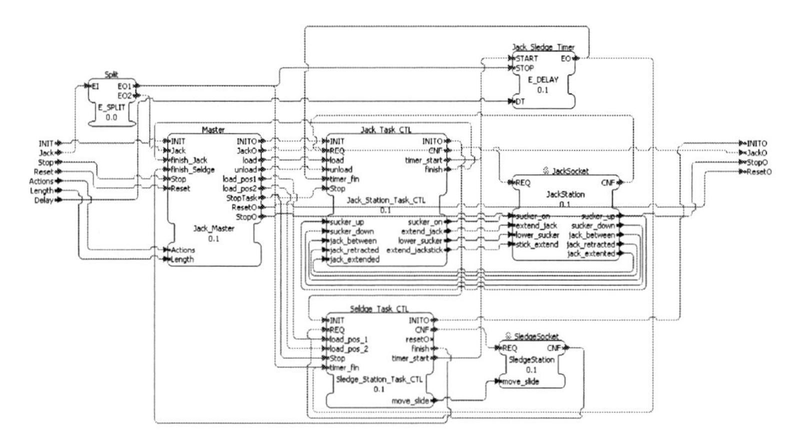

Figure 3.7 Parametrized Master- and Task-Controller of the Jack Station

There exist several possibilities to reconfigure the control application an therefore to change the production process. One approach is to develop for every new production scenario a new Master-Controller as described at the end of the last section, but therefore the used runtime environment has to support the dynamic reconfiguration as presented in [BVT+08]. Another way is to implement all possible production scenarios and switch between them through changed parameters.

Implementing all production scenarios with the first presented approach of a Central-Controller will not be suitable and even at the second approach with one Master- and several Task-Controllers, it would be quite challenging and maybe not possible to develop one Master-Controller for a plant part incorporating all production possibilities. Accordingly, the approach of distributed Master- and Task-Controller described in [MHH07] is extended. Thereby, every Master-Controller coordinates the actions of its underlying Task-Controllers with the other Master-Controllers of the control system. A lower number of coordinated

action	decrciption
0 - load	load a workpiece into the tin
1 - load_pos1	load a workpiece into the tin from the first sledge position
2 - load_pos2	load a workpiece into the tin from the second sledge position
3 - unload	unload a workpiece from the tin
4 - unload_pos1	unload a workpiece from the tin to the first sledge position
5 - unload_pos2	unload a workpiece from the tin to the second sledge position

Table 3.1 Possible scenarios of the Jack Station

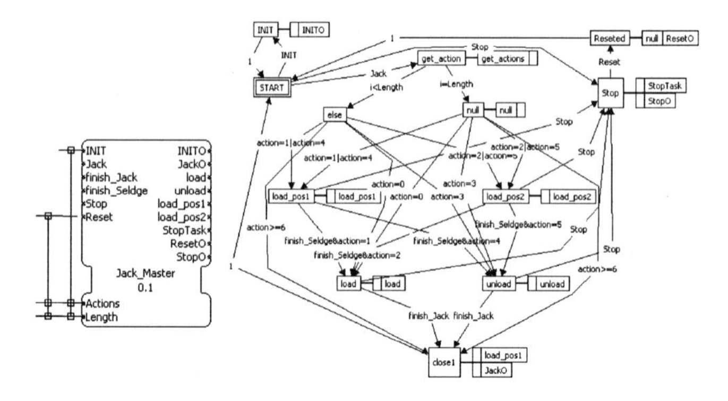

Figure 3.8 Master-Controller of the Jack Station

Task-Controller makes the development of the Master-Controller easier. At [GHH09] each conveyor as well as each mounted processing station gets its own Master-Controller. Each of them implements all possibilities to trigger the connected Task-Controllers. For example the Master-Controller of the Jack Station has to incorporate the scenarios of Table 3.1, if it controls the Task-Controller of the Slide Station as well as shown in Figure 3.7. The derived ECC is shown in Figure 3.8.

Through the data input *Actions* an array of sequentially performed actions can be parametrized. Each time the function block receives the input event *Jack* and the EC state *START* is active, the next action is read from the array and the connected Task-Controllers are triggered accordingly. If both Task-Controllers are ready, then the Master-Controller publishes the event *JackO*, to inform the other Master-Controllers to be ready and to release the pallet. If the number of performed actions equals the provided data value at *Length*, than the internal counter i will be set to zero again and the planed production scenario starts again. The resulting interconnection of the Master- and Task-Controllers of the processing station consisting of mechanical components Jack and Seldge Station is shown in Figure 3.7.

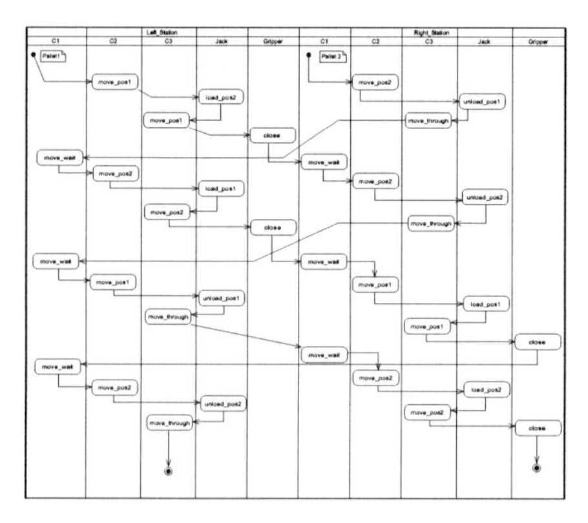

Figure 3.9 Activity Diagram for a production scenario with two pallets

As described in more detail in [GHH09], each production scenario can now be planned with an activity diagram of the Systems Modeling Language (SysML). According to Figure 3.9, a partition is inserted into the activity diagram for every Master-Controller, and each pallet is represented by concurrent activities. Each activity is mapped to the partition representing the Master- and Task-Controller performing the action, described by this activity. If the production scenario is ready, the array of actions for each Master-Controller can be read from top to down from the corresponding partition. These new action arrays can be send to the control devices through the available human machine interface, and the new production scenario is online.

3.5 Workpiece Controller

Increasing the number of pallets at the manufacturing system will increase the complexity of the planned production scenario and the planning will be the most time consuming part of the reconfiguration. Also if a pallet is added or removed from the plant or if there exist production alternatives, which should be exchanged by each other due to some reasons, the scenario has to be updated. Concerning flexibility and maintenance of the control application it is easier to develop the production scenario for every pallet separately. Each production scenario can be translated into a function block. This function block controls the actions the Task-Controller should perform. Therefore, it is named Workpiece Controller. Each Workpiece Controller allocates the necessary Task-Controllers for the following production step and release them afterwards. If not all necessary Task-Controllers are available and there exists a production alternative, then this one is chosen. Coming back to the idea of prioritising the EC transition at [TD06a], it is also possible to prioritise the production alternatives. At the presented testbed there exists for every pallet a Workpiece Controller and if now a new one is added to or removed from the manufacturing system the corresponding Workpiece Controller starts or ends to communicate with the Task-Controllers.

Figure 3.10 shows a function block network to control the Gripper Station, by sending the *RUN* event in combination with the event qualifier *store* and *take*. These event qualifiers allow the event propagation through the function block of the type *E_Permit* and due to their values the actions *close, hold* and *deposite* are triggered. The *Finish* event is

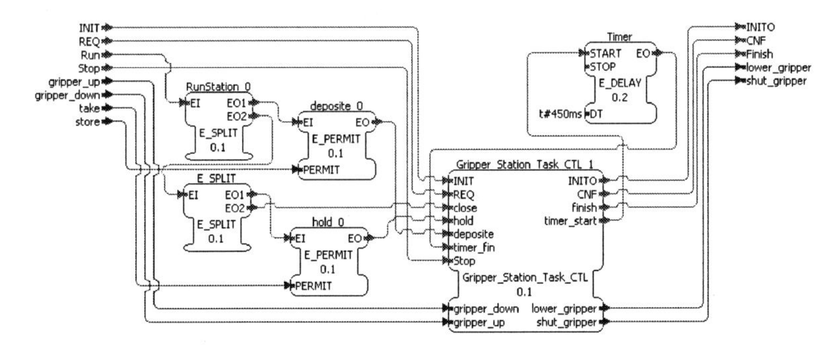

Figure 3.10 Function Block Network to control the Gripper Station with a bad coding practice

published if the performed action is accomplished. Any well-skilled IEC 61499 control engineer may realise the critical function block interconnection by splitting the events and using the *E_Permit* function block to interrupt an event chain. But this implementation results from the provided program specification:

1. perform the action *hold*, if the event *Run* occurs in combination with *take* = *true*,
2. perform the action *deposite*, if the event *Run* occurs in combination with *store* = *true* or
3. otherwise perform the action *close*.

During the ongoing work the former *W2-Function-Block-Controller (Western Reserve Controls)* and the *Netmaster II* controller from *Elsist S.l.r* were replaced by *Wago IPCs 750-860*. Thereby, the control applications are not any longer executed by the Java based FBRT runtime, but instead by the C++ based *Forte* runtime *0.3.7*. According to literature [SZC+06, Zoi09] this changes the scheduling of the function blocks from *common function calls*, where an event source immediately activates the sink and a kind of event propagation stack is realised, to a *first-in-first-out* queue. Thus, the event propagation between the function blocks changes, what is shown exemplarily for the input event *Run* and the associated data $take = true \wedge store = false$:

FBRT: Run → RunStation.EI; RunStation.EO1 → deposite.EI;
RunStation.EO2 → E_Split_0.EI; E_Split_0.EO1 → hold.EI;
hold.EO → Task_CTL.hold; ...

FORTE: Run → RunStation.EI; RunStation.EO1 → deposite.EI;
RunStation.EO2 → E_Split_0.EI; E_Split_0.EO1 → hold.EI;
E_Split_0.EO2 → Task_CTL.close; ...

This means porting the function block network from the FBRT to the FORTE runtime environment the plant behaviour changes from *lifting* a tin from a pallet, to *closing* a tin at a pallet. To correct this malfunction and to receive a truly portable function block network, cascaded *E_Switch* function blocks have to be used as presented in Figure 3.11 instead of the *E_Split* and *E_Permit* function blocks. This will change the event propagation for

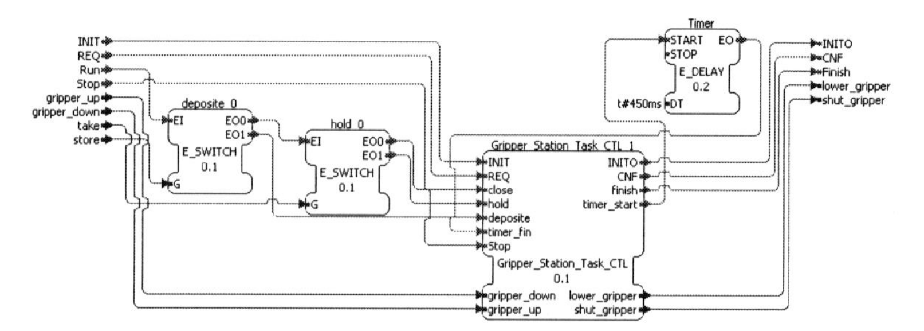

Figure 3.11 Function Block Network to control the Gripper Station with a better coding practice

both runtime environments to the following, if the same input event with the associated data is received.

FBRT/FORTE: Run → deposite.EI; deposite.EO0 → hold.EI;
hold.EO1 → Task_CTL.hold; ...

Coming back to all the different scheduling possibilities described in Section 1.1, it would be a hard and time consuming task for a control engineer to prove any possible event propagation, if there exists one or more not fulfilling the specifications of the plant behaviour. Thus, a formal model of the function blocks as well as of the scheduling is needed to prove the provided specifications.

3.6 Servo-Control System

Another common control example of industrial plants is a position control system, which could be found in disc drives, automotive products, robotics, process control and gantry cranes. A simple position control system is shown in Figure 3.12a and consists of a servo motor, an optical or magnetic position sensor and a controller. Thereby, the shaft position is detected by the position sensor and expressed in a 8 bit Gray code, depending on the desired solution precision. This signal will be decoded by the controller first and then subtracted from the provided reference position. Thus, the output of the 8 bit adder in subtract mode will be the position error, which is the input to an *or* logic to decide whether the servo motor should be turned on or off and in which direction. A more detailed description concerning

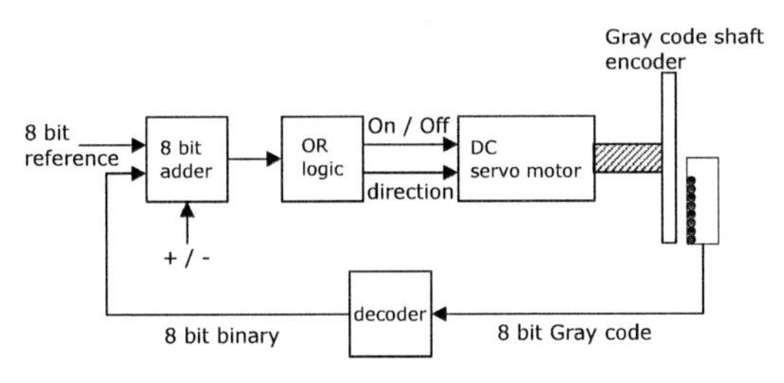

(a) CE300 position control system

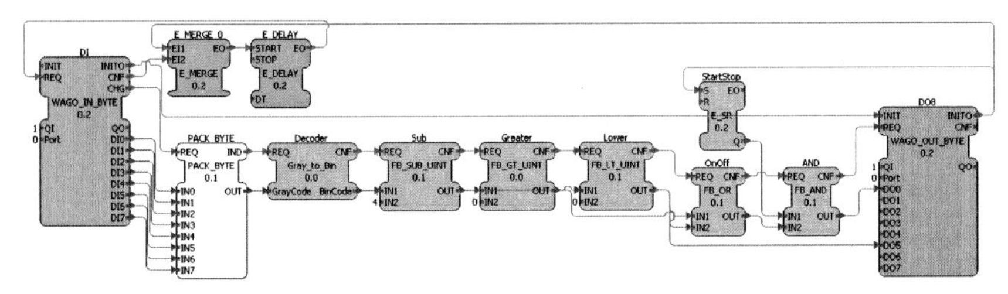

(b) Control Application of the Servo-Control System

Figure 3.12 Servo-Control System

the control and simulation of such a servo control system could be found in [Rea05].
The challenges of the presented control systems lie in the complexity of the execution behaviour and the data processing. At the distributed control system of the manufacturing plant, a hierarchical control structure with several Master- and Task-Controllers is used. This leads to a complex execution behaviour, but to a simple data processing with mainly Boolean values. Contrary to this, the function block network of the servo control system in Figure 3.12b has a linear execution behaviour from the decoder, to the adder and finally to the logic, but a complex data processing with integer-valued data. Both control systems are implemented with function blocks following the IEC 61499 and should be verified using a closed-loop model to ensure the correct plant behaviour. Therefore, a formal model divided into an execution and a data processing model of such function blocks as well as a scheduling model for possible runtime implementation will be described in the following chapter.

3.7 Summary

The chapter starts with a brief description of the used testbed in this thesis and provides a short overview about the software model and the used elements defined by the 1st part of IEC 61499 standard. Starting with approach of the Central Controller, where one monolithic function block controls the whole plant, several control approaches are presented to modularise the control implementation and meet the introductorily claimed key issues of modern automation systems. All of the going presented approaches have in common to use the same Task-Controllers. Each Task-Controller implements all elementary actions the corresponding mechanical component can perform. Thus, each vendor of mechanical components can supply the Task-Controllers as his intellectual property and encapsulate it inside a function block [VCL05]. Due to the defined requirements for software tools at the [IEC-61499-2], it is ensured, that every engineering environment can import these vendor-specific function blocks.
During the replacement of the former *Netmaster II* and *W2-Function-Block-Controller* (Western Reserve Controls) by *Wago IPCs 750-860*, the used runtime environment changed from the Java based FBRT to the C++ based Forte. The function block network remained untouched, but the execution behaviour changed to performing the action *close* every time the input event REQ occurs, without the evaluation of event qualifiers. The reason for this lies at the different *scheduling functions* as mentioned in Section 1.1. But how can a hardware vendor developing function blocks or a control engineer forced to redistribute a control application ensure the same execution behaviour for every runtime environment?
Therefore, the following chapters will present a solution based on model checking techniques of the closed-loop system incorporating a formal model of the plant and the controller as well. To develop in a semi-automatical way the formal controller model, transformation rules for the distributed control system into the used formal modelling language *discrete timed Net Condition/Event Systems*presented in Chapter 2 are defined in the following chapter. Chapter 5 will present the modelling of the closed-loop system, starting with the transformation of the function blocks and interconnecting them to each other as well as to the process and communication interface to gain the model of a control device. At the end of Chapter 5, the device model is connected to the plant model. The verification and scheduling results for the presented control examples are presented in Chapter 6.

Chapter 4

Formal Modelling of the IEC 61499 elements

During the last couple of years the author specified several transformation rules of function blocks and their interconnection to function block networks into the formal modelling language *Net Condition/Event Systems*. These transformation rules will be summarized, rearranged and extended in this chapter. Thereby, they are grouped into the *execution, data processing* and *scheduling model*. The general transformation rules of the execution model are presented in [IVGH07] and extended to the use of integer-valued data types at [IVGH08]. Rules to transform the data processing inside the algorithms of the basic function blocks are defined in [GIVH08, GIVH10]. In [GH10], the execution model was extended once again, to connect it to the scheduling model of different runtime environments.
Due to specification of the graphical appearance of the NCE module in [Kar09], the first transformation rule concerning the graphical representation gets obsolete. Thus, the interface of an NCE module is represented similarily to the one used for function blocks.

4.1 Execution Model

As mentioned at the introduction of the IEC 61499 software model in Section 3.1, at the execution model of function blocks it has to be distinguished between composite, basic and simple function blocks. All of them have an interface consisting of events and data in- and outputs. Each data input is sampled if an associated event occurs and each data output is published if an associated event is published. Therefore, at least one or even more event inputs (outputs) should be associated with a data input(output). Otherwise the set of associated data inputs (outputs) to an event input (output) may be empty. The interface is the only common part of all function blocks. A simple function block incorporates only algorithms, which are triggered if an input event with the same name occurs. At basic function blocks the execution of algorithms and the publishing of output events and data is controlled by the Execution Control Chart and composite function blocks have a function block network inside.
In the following, rules will be presented to transform this function block elements into the formal modelling language and to provide the connections to the data processing as well as to the scheduling model.

Transformation rule 1 - Graphical Representation: This rule is obsolete since the publication [Kar09].

Transformation rule 2 - Interface: During the transformation process of a function block each NCE module gets the condition inputs *Resource_Idle* and *Enable_EvOutputs*, the event outputs *FB_scheduled* and *FB_exec_finished* as well as the condition output *FB_idle*.
At the transformation of a

simple or basic function block the place invariant *FB_idle – FB_scheduled* is inserted first with an initial marking of one token at the place *FB_ilde.* The connecting transitions have the event mode $\boxed{\vee}$. The transition at the pre-set of *FB_Idle* gets the name *FB_exec_finished* and has an event output arc to *FB_exec_finished.* The other one gets the name *FB_scheduled* and is connected to the event output *FB_scheduled.* Place *FB_idle* is connected to the condition output *FB_idle* by a condition output arc.
Afterwards, rule 2.1 is applied to every data input and rule 2.2 to every data output. Next the transformation rule 2.3 is applied to every event input, and additionally each transition **_Release* gets a condition input arc from the condition input *Resource_Idle.* Finally, transformation rule 2.4 is used to transform every event output, and each transition **_Release* is connected by a condition input arc to the condition input *Enable_EvOutputs.*

composite function block *X* two additionally NCE modules named *X_inputs* and *X_outputs* are created. The NCE structures received by applying the transformation rules 2.1 and 2.3 are inserted into the first and the other into the second module. Afterwards, the input interface of the module *X_inputs* is copied to the outputs of the same module. Thereby, each event output * gets an event output arc from the corresponding transition **_Release*, and every condition output + has a condition output arc from the places *+_True*.
Finally, the output interface of the module *X_outputs* is copied to the input interface of the same module. Thereby, all event inputs * get an event input arc to the

(a) Place invariant FB_idle – FB_scheduled (b) Transformed Interface of the composite FB Gripper_Reuse

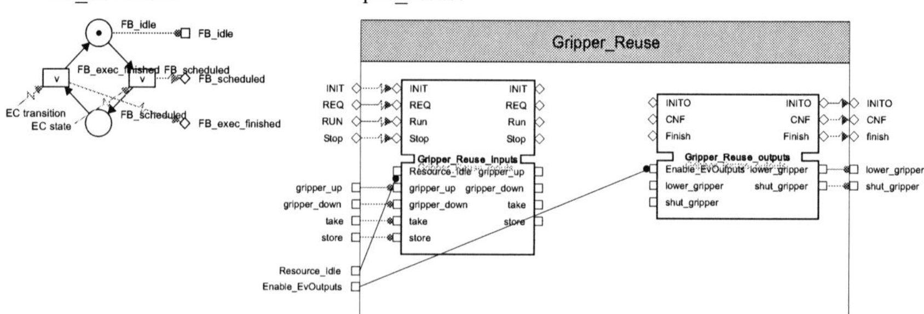

Figure 4.1 Transformation rule 2 - Interface

transition *_*Set* and every condition input + is connected by a condition input arc to the transition +_*toTrue* and by an inhibitor input arc to the transition +_*toFalse*. The last transformation step for the interface of a composite function block is to copy the input interface of the NCE module named *X_inputs* to the input interface of the transformed NCE module and connect both by event and condition interconnections. Also the output interface of the NCE module named *X_outputs* is copied to the output interface of the transformed NCE module and both interfaces are connected by event and condition interconnections.

Figure 4.1a shows the created place invariant during the transformation of a basic function block and its connection to interface. During the transformation of the EC transitions and EC states, several event arcs will be connected to the transitions as shown. Applying the rule above to the interface of the composite function block *Gripper_Reuse* presented in Figure 3.10 at Page 37, the composite NCE module of Figure 4.1b will be created. It consists of the base NCE module *Gripper_Reuse_inputs* and *Gripper_Reuse_outputs* as well as the event and condition interconnections to the interface.

Transformation rule 2.1 - Data Input: At first the data type of the data input to-be-transformed is checked. If it is a Boolean data input, transformation rule 2.1.1 is applied, and if it is an integer-valued one, rule 2.1.2 has to be used. If the data input has an array as data type, the last rule 2.1.3 is used for the transformation process.

Transformation rule 2.1.1 - Boolean Data Input: Each Boolean data input + of a function block is transformed to a condition input + of an NCE module. This condition input is extended by an NCE structure consisting of a place invariant and its connecting transitions. The places get the names +_*True* and +_*False*, and the transitions are named +_*toTrue* and +_*toFalse*. Both transitions have the event mode $\boxed{\vee}$. Furthermore, a condition input arc is connected from the condition input to the transition +_*toTrue*, and an inhibitor input arc is connected to the transition +_*toFalse*.

Transformation rule 2.1.2 - Integer-valued Data Input: For each integer-valued data input + of a function block, the number n of representing bits have to be retrieved first. Next, the place + is inserted with the capacity of $2^n - 1$. Afterwards, rule 2.1.1 is repeated n times with the iterator i and providing the string +_*2pi*. During this iteration, flow arcs with the arc weight 2^i are inserted from the transition +_*2pi_toTrue* to the place + as well as from the place + to the transition +_*2pi_toFalse*.

Transformation rule 2.1.3 - Data Input with an array: For each data input + of a function block with an array as data type, the elementary data type of the array has to be retrieved first. Next, each array position is transformed by applying rule 2.1.1 or 2.1.2 depending on the elementary data type and providing the string +_*i*.

Applying the transformation rule 2.1 and its sub-rules to the Boolean data input *gripper_down* of a basic function block, the NCE structure and In- and Output structure in Figure 4.2a is created. The transformation result for the 8 bit integer-valued data input

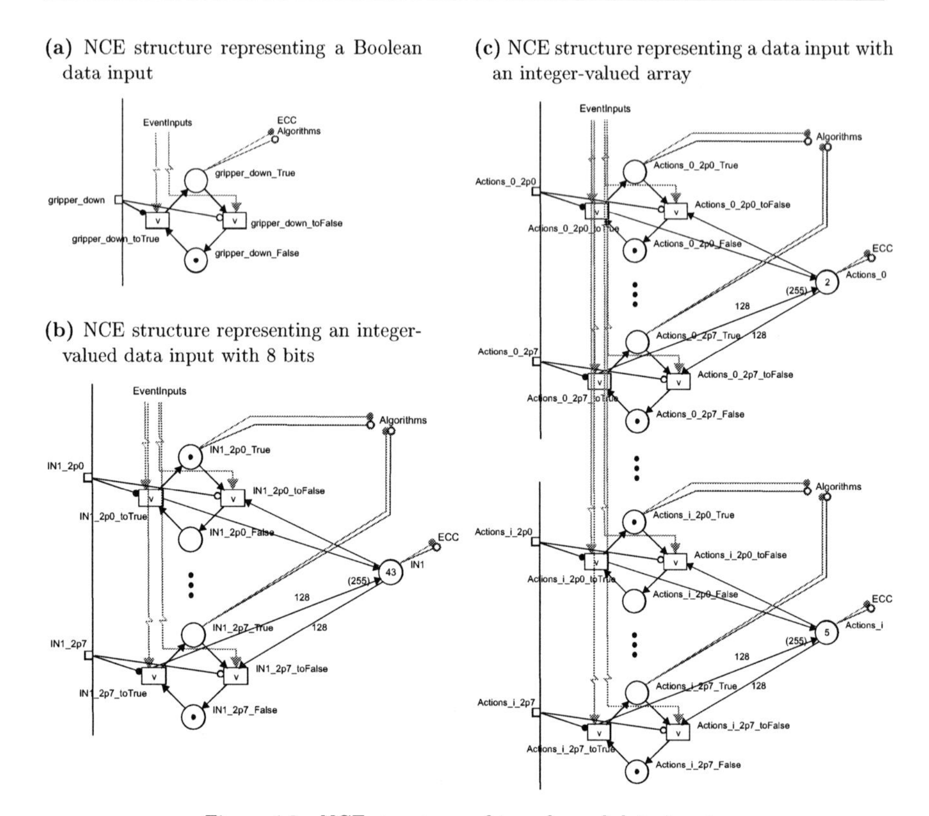

(a) NCE structure representing a Boolean data input

(b) NCE structure representing an integer-valued data input with 8 bits

(c) NCE structure representing a data input with an integer-valued array

Figure 4.2 NCE structures of transformed data inputs

action of a basic function block is shown in Figure 4.2b. If the transformation rule is applied to the data input *Actions* with an integer-valued array, the NCE structure and In- and Output structure in Figure 4.2c will be the result.

Transformation rule 2.2 - Data Output: At first, the data type of the to-be-transformed data output is checked. If it is a Boolean data output, transformation rule 2.2.1 is applied, and if it is an integer-valued one, rule 2.2.2 has to be used. If the data output has an array as data type, the last rule 2.2.3 is used for the transformation process.

Transformation rule 2.2.1 - Boolean Data Output: Each Boolean data output $+$ of a function block is transformed to a condition output $+$ of an NCE module. This condition output is extended by an NCE structure consisting of two place invariants and its connecting transitions. The first invariant is a buffer for internal value changing by any algorithm. The places get names *+_Buffer_True* and *+_Buffer_False*, and the transitions are named *+_Buffer_toTrue* and *+_Buffer _toFalse*. Both transitions have the event mode [V]. The second invariant represents the actual published value, and therefore the places get the names *+_True* and *+_False* and the transitions are named *+_toTrue* and

(a) NCE structure representing a Boolean data output (the actual buffer value is published)

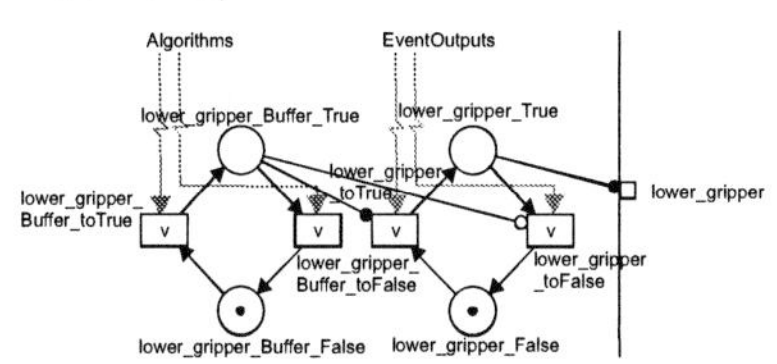

(b) NCE structure representing an integer-valued data output with 8 bits (the actual buffer value is published)

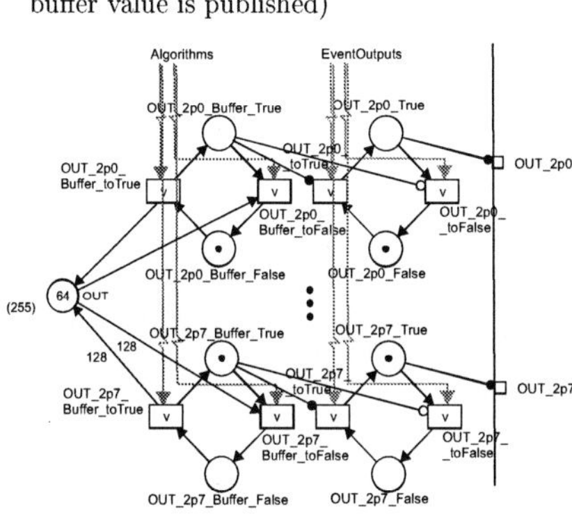

(c) NCE structure representing a data input with an integer-valued array (the actual buffer value is unpublished)

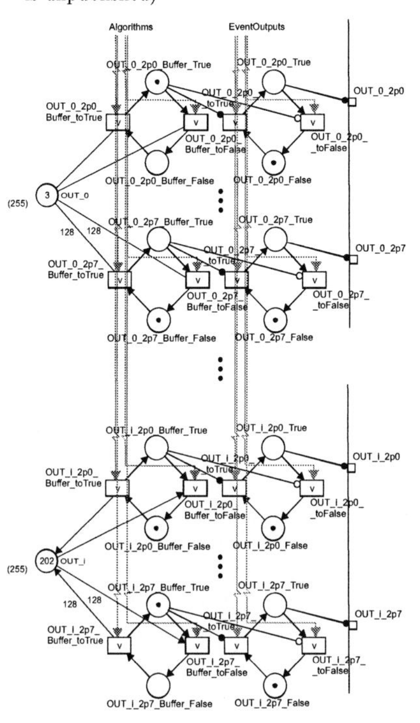

Figure 4.3 NCE structures of transformed data outputs

+_ toFalse. Once again, the transitions get the event mode $\boxed{\vee}$. Furthermore, a condition arc is connected from the place *+_ Buffer_ True* to the transition *+_ toTrue* as well as an inhibitor arc to the transition *+_ toFalse.* Finally, a condition output arc is inserted to connect the place *+_ True* and the condition output.

Transformation rule 2.2.2 - Integer-valued Data Output: For each integer-valued data output *+* of a function block, the number n of representing bits has to be retrieved first. Next, the place *+* is inserted with the capacity of $2^n - 1$. Afterwards, rule 2.2.1 is repeated n times with the iterator i and providing the string *+_ 2pi.* During this iteration, flow arcs with the arc weight 2^i are inserted from the transition *+_ 2pi_ Buffer_ toTrue* to the place *+* as well as from the place *+* to the transition *+_ 2pi_ Buffer_ toFalse.*

Transformation rule 2.2.2 - Data Output with an array: For each data output *+* of a function block with an array as data type, the elementary data type of the array has to be retrieved first. Next, each array position is transformed by applying rule 2.2.1 or 2.2.2 depending on the elementary data type and providing the string *+_ i.*

Figure 4.3a shows the transformation result for the Boolean data output *lower_gripper* of a basic function block, and in Figure 4.3b the transformation rule 2.2 is applied to the 8 bit integer-valued data output *OUT* of a basic function block. If the same transformation rule is applied to a data output *OUT* with an integer-valued array, the NCE structure and In- and Output structure in Figure 4.3c is the result.

Transformation rule 2.3 - Event Input Each event input * of a function block is transformed to an event input of the corresponding NCE module . This event input is extended by an NCE structure consisting of a place invariant and its connecting transitions. The places get the names **_Set* and **_Released*, and the transitions are named **_Set* and **_Release*. The transitions **_Release* get the firing mode *instantaneous*, which means it has to be fired before any spontaneous transition. Furthermore, an event input arc is connected from the event input to the transition **_Set*.
For every associated data input to an event input, event arcs have to be inserted with their source at **_Set* and their sinks at *+_toTrue* and *+_to False*, where + stands for the name of the associated data input and is extended by *_2pi*, *_j* or *_j_2pi* depending on the data type. This association is represented by the attribute *Var* of the *With* tag, which is a child element of the *Event* tag.

Transformation rule 2.4 - Event Output: Each event output * of a function block is transformed to an event output of an NCE module. This event output is extended by an NCE structure consisting of a place invariant and their connecting transitions. The places get the names **_Set* and **_Released* and the transitions are named **_Set* and **_Release*. The transition **_Release* has the firing mode *instantaneous*, and the transition **_Set* has the event mode [V]. Furthermore, an event output arc connects the transition **_Release* and the event output.
For every associated data output to an event input, event arcs have to be inserted with their source at **_Set* and their sinks at *+_toTrue* and *+_to False*. Depending on the data type of the associated data output, the + stands for the name only or is extended by *_2pi*, *_j* or *_j_2pi*. Each association is represented by the attribute *Var* of the *With* tag, which is a child element of the *Event* tag.

Applying the transformation rule 2.3 to the event input *REQ* of a basic function block

(a) NCE structure representing an event input **(b)** NCE structure representing an event output

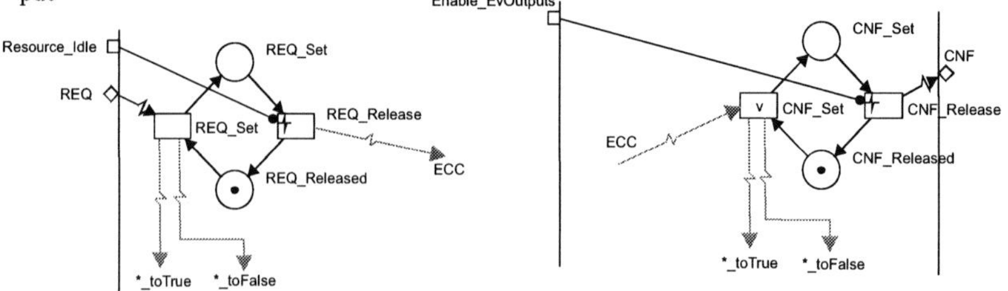

Figure 4.4 NCE structure of a transformed event in- and output

the NCE structure and In- and Output struture in Figure 4.4a is the result. Figure 4.4b presents the transformation results of the event output *CNF*.

After transforming the interface of a function block, the internal elements have to be transformed. A simple function block has only algorithms as additional elements, and their execution is triggered by an input event with the same name as the algorithm. At basic function blocks, the execution of the algorithms is controlled by the Execution Control Chart (ECC). The ECC consists of EC states and EC transitions to connect them. Every time an EC transitions clears, the following EC state is activated, and the algorithms of the associated EC actions are executed. During the execution of the algorithms, no following EC transition is allowed to clear. After the completion of all algorithms, the output events of all associated EC actions are generated and published sequentially as well as the following EC transitions are evaluated.
Each composite function block incorporates a function block network. The event and data inputs are connected to several component function blocks of the function block network. To each event or data output, one event or data output of a component function block may be connected. But the transformation of the function block network has to incorporate also the scheduling of the component function blocks. Due to this manner of fact, the transformation of the function block network is part of the scheduling model in Section 4.3.

Transformation rule 3 - Execution Control Chart:

Transformation rule 3.1 - EC state: Each EC state *$* with no associated EC action is transformed to the place *$_Idle*. Each EC state *$* with at least one associated EC action is transformed to the three places *$_sched_Algs*, *$_waiting_Algs* and *$_Idle*. The transition between the first and second place is named *$_sched_Algs*, and the other one between the second and third place has the name *$_Algs_complete*. From the transition *$_Algs_complete*, an event arc is connected to the transition *FB_exec_finished*, created during the interface transformation.
If the EC state is the initial state, then the place *$_Idle* gets the initial marking of one token.
Each associated EC action is transformed by applying rule 3.2.

Transformation rule 3.2 - EC action: Each EC action may consist of an algorithm to be executed and an output event to be published. The algorithm *AlgName* to be executed is transformed to an event arc connecting the transition *$_sched_Algs* of the EC state *$* the EC action is associated to and the transition named *AlgName_toRun* as well as a condition arc from the place *AlgName_Wait* to the transition *$_Algs_complete* of the EC state the EC action is associated to. Each output event to be published by the EC action is transformed to an event arc connecting the transition *$_Algs_complete* of the EC state the EC action is associated to and the transition **_Set* of the corresponding event output.

Applying the transformation rules 1, 2 and 3 to the function block *E_REND* defined the the appendix A of the standard, the NCE module in Figure 4.5c is created. At the left side of the module the transformed event inputs are located and at the right side the event output *EO*. Furthermore, the place invariant *FB_idle – FB_scheduled* is located at the right. In the middle the transformed ECC of Figure 4.5b can be found.

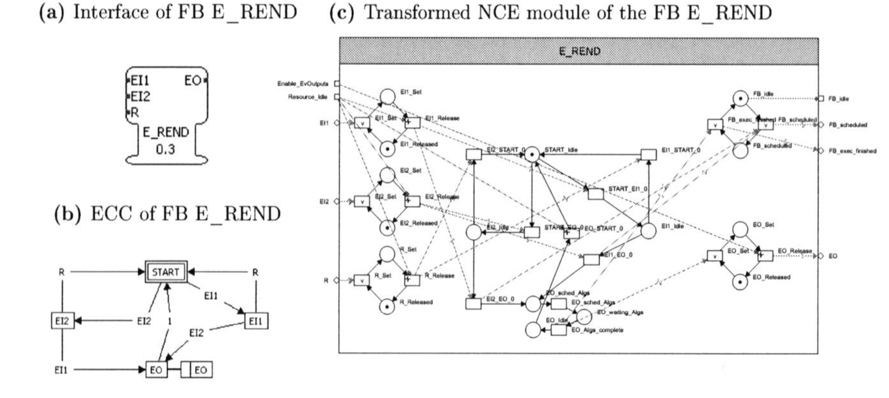

(a) Interface of FB E_REND (c) Transformed NCE module of the FB E_REND

(b) ECC of FB E_REND

Figure 4.5 Transformation of the function block E_REND

Due to the fact, the EC states *START, EI1* and *EI2* do not have an associated EC action, they are only transformed to the places *START_Idle, EI1_Idle* and *EI2_Idle.* Despite this, the EC state EO is transformed to the places *EO_sched_Algs, EO_waiting_Algs* and *EO_Idle* as well as to the connecting transitions *EO_sched_Algs* and *EO_Algs_Complete.* As initial state of the ECC the EC state START is marked and due to this the place START_Idle holds the initial marking of one token. The EC transitions connecting the EC states are transformed according to rule 3.3 defined in the following.
As described in [DV08, CA08], the condition of an EC transition is a Boolean equation consisting of zero or one event input variable and a set of data inputs and internal variables. The layout of the equation is as follows. At first, an event input variable may occur and following an equation using the data inputs and internal variables. To describe the transformation of an EC transition, the following example should be taken into account. An EC transition connects the EC states X and Y and has the condition

$$Jack \wedge (action = 3 \vee (action < 6 \wedge sucker\_up)).$$

At first the brackets have to be resolved by the use of the distribution law. This leads to the disjunctive term

$$Jack \wedge action = 3 \vee Jack \wedge action < 6 \wedge sucker\_up.$$

Each conjunctive term of this disjunction is modelled by a transition between the places *X_Idle* and *Y_sched_Algs* and having the transition *Jack_Release* as event source. Additionally, condition and inhibitor arcs to the places *action* and *sucker_up_true* are inserted. Due to the fact that there are two conjunctive terms, one transition is named *X_Y_0* and the other *X_Y_1.*

Transformation rule 3.3 - EC transition: By using the law of distribution and de Morgan, all brackets of a condition assigned to the ECTranstion have to be resolved to get n conjunctive terms connected through disjunctions. If the EC state at the post-set

of the ECTranstion has at least one EC action assigned, each conjunctive term has to be modelled by a transition that is between the *§_Idle* place of the previous EC state and the *$_sched_Algs* place of the following EC state with the name as concatenation of the names of both EC states and the string *_n*. Furthermore, an event arc has to connect the actual transition with the transition *FB_scheduled.*
If the EC state at the post-set of the ECTranstion has no EC action assigned, then the post-arc of the inserted transition has to be connected to the place *$_Idle*, and no event arc to the transition *FB_scheduled* is inserted.
If the first variable of a conjunctive term is an event, an event arc from the transition **_Release* to the actual transition has to be inserted. Otherwise, the firing mode of the actual transition has to be set to *instantaneous*, and a condition input arc has to be connected to the condition input *Resource_Idle*. The remaining part of the term represents a Boolean equation of internal variables and data inputs, which can be modelled with condition and inhibitor arcs connecting the place *+_True* and the actual transition. If there is a comparison done between a variable X with integer-valued data type and a defined value i, then rule 3.3.1 up to 3.3.6 has to be taken into account.

Coming back to the transformation of the function block E_REND presented in Figure 4.5, the transformed EC transitions with the sink *START, EI1* or *EI2* are always connected at the post-set to the place *$_Idle* and all of them have no outgoing event arc. Despite this, the transformed EC transition with the sink *EO* have at the post-set the place *EO_sched_Algs*. Furthermore, only these transitions have an outgoing event arc to the transition FB_scheduled. Since, the EC transition between the EC state *EO* and *START* has no input event used at the condition, the corresponding transition at the formal model has the firing mode *instantaneous*.

Transformation rule 3.3.1 - $X > i$: In order that X should be greater than i, a condition arc with arc weight *i+1* has to be connected to the transition with its source at place X.

Transformation rule 3.3.2 - $X \geq i$: In order that X should be greater or equal to i, a condition arc with arc weight i has to be connected to the transition with its source at place X.

Transformation rule 3.3.3 - $X < i$: In order that X should be lower than i, an inhibitor arc with the arc weight i has to be connected to the transition with its source at place X.

Transformation rule 3.3.4 - $X \leq i$: In order that X should be lower or equal to i, an inhibitor arc with the arc weight *i+1* has to be connected to the transition with its source at place X.

Transformation rule 3.3.5 - $X = i$: In order that X should be equal to i, a condition arc with the arc weight i and an inhibitor arc with the arc weight *i+1* have to be connected to the transition with their sources at place X.

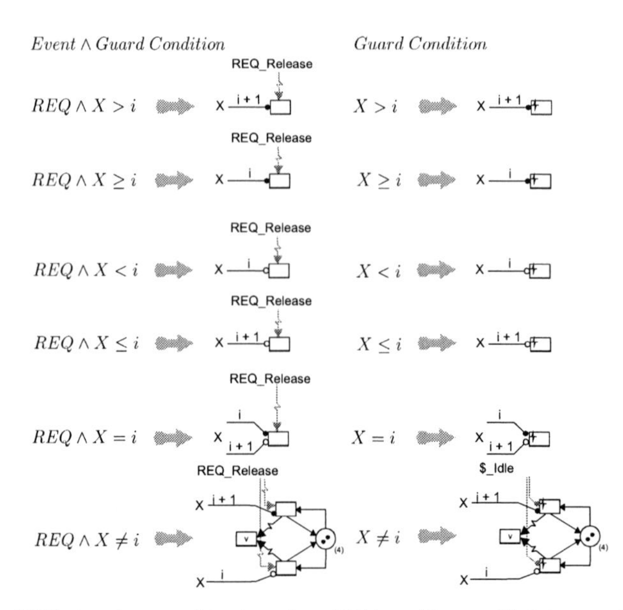

Figure 4.6 NCE structure used to transform EC transitions with integer-valued data at the guard condition

Transformation rule 3.3.6 - $X \neq i$**:** In order that X should not be i, transformation rule 3.3.1 and 3.3.3 have to be combined. Thus, an inhibitor arc with the arc weight i has to be connected to one transition and a condition arc with arc weight $i+1$ to a second transition. Both arcs have as their source place X. The two transitions are part of the pre- and post-set of a place with 2 tokens and the capacity of 4. Furthermore, they are connected via event arcs to the resulting transition with the event mode $\boxed{\vee}$.

The transformation of algorithms is divided into a general and a conditional part. The conditional part represents the data processing model of the following section, and the general part is an NCE structure connecting the execution and the data processing model.

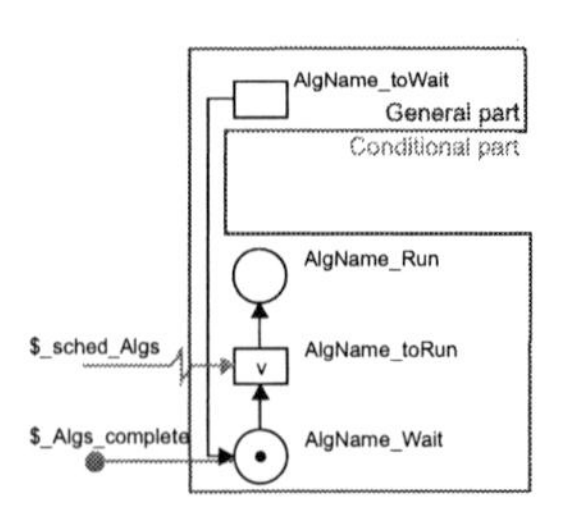

Figure 4.7 NCE structure of the general part of the algorithm transformation

Transformation rule 4 - Algorithm As shown in Figure 4.7, each algorithm is transformed to an NCE structure consisting of a marked place *AlgName_Wait* and the transition *AlgName_toWait* at the pre-set and the transition *AlgName_toRun* at the post-set. The transition *AlgName_toRun* has the event mode $\boxed{\vee}$and the place *AlgName_Run* at post-set.

Afterwards, transformation rule 4.1 to create the conditional part is applied to the algorithm. Concluding, a flow arc is inserted from the place *AlgName_Ln_Fin* of the last transformed algorithm line to the transition *AlgName_toWait.*

4.2 Data Processing Model

All ongoing explanations to the data processing model are written in a way to be easily adopted to the previously defined execution model of function blocks, but the presented data processing model can be used also with any other execution models as long as the formal model is *Net Condition/Event Systems* and a binary representation of integer-valued data is chosen. Thus, it is possible to use the algorithm transformation rules at the end of this section for transforming controllers with a different execution runtime than the one of the IEC 61499 [GPH10]. Most of the approaches presented at the introduction and including data processing are limited to Boolean data processing, but the integer-valued data may be processed by adding, subtracting, multiplying or comparing two values as well. As known from [SS04, BDM05], the base operation of all four mathematical operations is the addition of n-digit binary numbers, which have to be realized first. Instead of modelling the subtraction of one value from another, the negated value of the second one and an additional carry bit could be added. In almost the same manner the models of a multiplier or divider can be derived from proven methods of informatics.
In [HM98] a Petri net model of *Carry-Ripple-Adder* is proposed and applied to the verification of instruction list programs. Thereby, each variable is modelled in a binary form, where each bit is represented by two places forming a place invariant. A similar approach is used for transforming the function block data inputs (outputs) and internal variables of integer-valued data types at the previous section. As could be seen at the function block network of Figure 3.12b at Page 39, most of the function blocks processing integer values have the data inputs *IN1* and *IN2* as well as the data output *OUT*. Thus, the examples presented at the following use the same. According to the execution model, the data inputs are represented as place invariants of the places *IN1_2pi_True* and *IN1_2pi_False* as well as *IN2_2pi_True* and *IN2_2pi_False*. Furthermore, the variable to store the result is the buffer of *OUT* and will be changed during the internal data processing, which is done by forcing the transitions *OUT_Buffer_2pi_toTrue* and *OUT_Buffer_2pi_toFalse*. By occurrence of the associated output event, the actual buffer value will be published.

4.2.1 Carry-Ripple-Adder

In a first sketch, the modelling of a *Carry-Ripple-Adder*, also known as *Carry-Chain-Adder* will be presented. The *Carry-Ripple-Adder* calculates the sum OUT_i and the carry c_i from the lowest to the highest digit i as follows ($\oplus \ldots XOR$):

$$\begin{aligned} OUT_i &= IN1_i \oplus IN2_i \oplus c_{i-1} \\ c_i &= IN1_i\, IN2_i \vee c_{i-1}\, (IN1_i \oplus IN2_i) \end{aligned}$$

Because of calculating bit by bit, the result and the carry for the next bit, n steps have to be fired at the NCE structure of the *Carry-Ripple-Adder*. Each step of the addition is forced from the algorithm by an event shown red inside figure 4.8b. Thereby, only one transition of the modelled addition step i will be condition-enabled to switch the bit *2pi* of the modelled buffer of the data output *OUT* to true or false. The conditions modelled at the forced transitions are done according to the truth table 4.8a by condition and inhibitor arcs connected to the place *IN1_2pi_True* or *IN2_2pi_True*. Inside the presented NCE

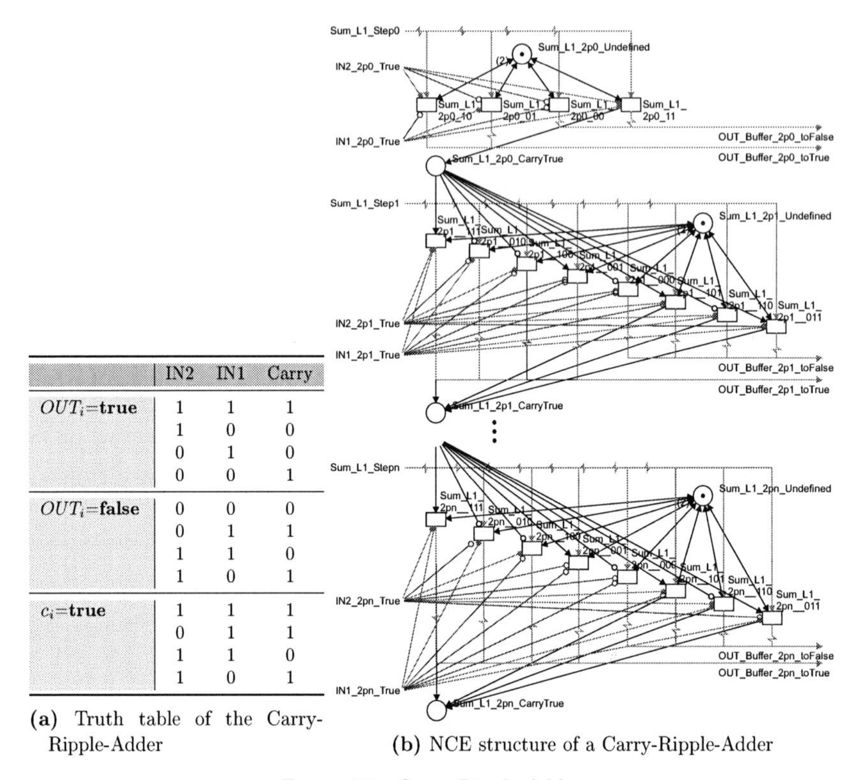

	IN2	IN1	Carry
OUT_i=**true**	1	1	1
	1	0	0
	0	1	0
	0	0	1
OUT_i=**false**	0	0	0
	0	1	1
	1	1	0
	1	0	1
c_i=**true**	1	1	1
	0	1	1
	1	1	0
	1	0	1

(a) Truth table of the Carry-Ripple-Adder

(b) NCE structure of a Carry-Ripple-Adder

Figure 4.8 Carry-Ripple-Adder

structure, the left 4 or at the first step the left 2 transitions are representing the cases where the result of *OUT_Buffer_2pi* gets *true* and the others switch the result to *false*.

4.2.2 Carry-Lookahead-Adder

One point to cope with the state space explosion problem during the reachability calculation is the use of efficient NCE structures. In contrast to the presented adder in [PV07] with an exponential number of fired steps ($O(2^n)$), the presented NCE structure of a *Carry-Ripple-Adder* with a linear number of fired steps ($O(n)$) is better. Also, it is now possible to add two variables and not only a static value to a variable. But concerning the methods of informatics, it is possible to reduce the linear number of fired steps to a logarithmic one, by deriving the NCE structure of a *Carry-Lookahead-Adder*. The main idea of Ladner and Fischer was to improve the calculation of the carry bits [LF80]. Thereby, the attribute *generate* $g_{i,j}$ or *propagate* $p_{i,j}$ is evaluated for each digit block *[i,j]* with $j \leq i < n$.
To ease the ongoing explanations the symbol † stands everywhere for the algorithm name and the line number.

First, these attributes are evaluated for the block $j = i$ as follows.

$$
\begin{aligned}
p_{i,i} &= IN1_i \oplus IN2_i \\
g_{i,i} &= IN1_i \, IN2_i
\end{aligned}
$$

Afterwards, the attributes of the blocks *[i,k+1]* and *[k,j]* will be merged until $j = 0$, by the following rules.

$$
\begin{aligned}
p_{i,j} &= p_{i,k+1} \, p_{k,j} \\
g_{i,j} &= g_{i,k+1} \vee p_{i,k+1} \, g_{k,j}
\end{aligned}
$$

The attribute *generate* is left hand stable, which means if the most significant block *[i,j]* has the attribute *generate*, then the block *[i,0]* has it also. This means, inside the NCES structure there will be only a place named †_gi0, and any transition evaluating the *generate* attribute to *true* will have a post-arc to this place. Thus, also the transitions named †_11_g_ii evaluating the *generate* attribute at the first step by checking the digit i of variable $IN1$ and $IN2$.
As shown at *Step i* in Figure 4.9 the *propagate* attribute is modelled by the place †_p_ij at every evaluation step. Only at the first step this place has the capacity of two.
By merging the blocks *[i,k+1]* and *[k,j]* the token at the place †_p_kj or †_g_kj will condition enable only one transition, which switches the state of the modelled attribute of the new block *[i,j]* to *propagate* or *generate*. During the evaluation process, the n tokens of place †_add flow directly to the place †_g_i0 or through place †_p_ii to †_g_i0 or to †_p_i0.
Using this information, the resulting sum of digit i is:

$$
OUT_i = p_{i,i} \oplus (g_{i-1,0} \vee p_{i-1,0} \, c_{-1}).
$$

Due to the modelling of an adder, the input carry c_{-1} will be zero, which reduces the formula above to $OUT_i = p_{i,i} \oplus g_{i-1,0}$ and the resulting truth table to Table 4.1.

OUT_i	$p_{i,i}$		$g_{i-1,0}$	
0	0	$\multimap$	0	$p_{i-1,0} \rightarrow$
1	1	$\rightarrow$	0	$p_{i-1,0} \, g_{i-1,0} \multimap$ $p_{i-1,0} \rightarrow$
0	1	$\rightarrow$	1	$g_{i-1,0} \rightarrow$
1	0	$\multimap$	1	$g_{i-1,0} \rightarrow$

Table 4.1 Truth table of the Carry-Lookahead-Adder

According to the number of rows, the modelling with an NCE structure is done by five different transitions with inhibitor and flow arcs from place †_p_ii and flow arcs from place †_p_i-10 or †_g_i-10. Only the transition representing the second row has two inhibitor arcs to the mentioned places. This means the block *[i-1,0]* absorbs any carry. The names of the transitions are †_ag_p_2pi, †_p_a_2pi, †_p_p_2pi, †_p_g_2pi and †_ag_g_2pi. The place †_add is at the post-set of all transitions. Thus, the number of tokens there will be n at the start of the calculation and at the end again.

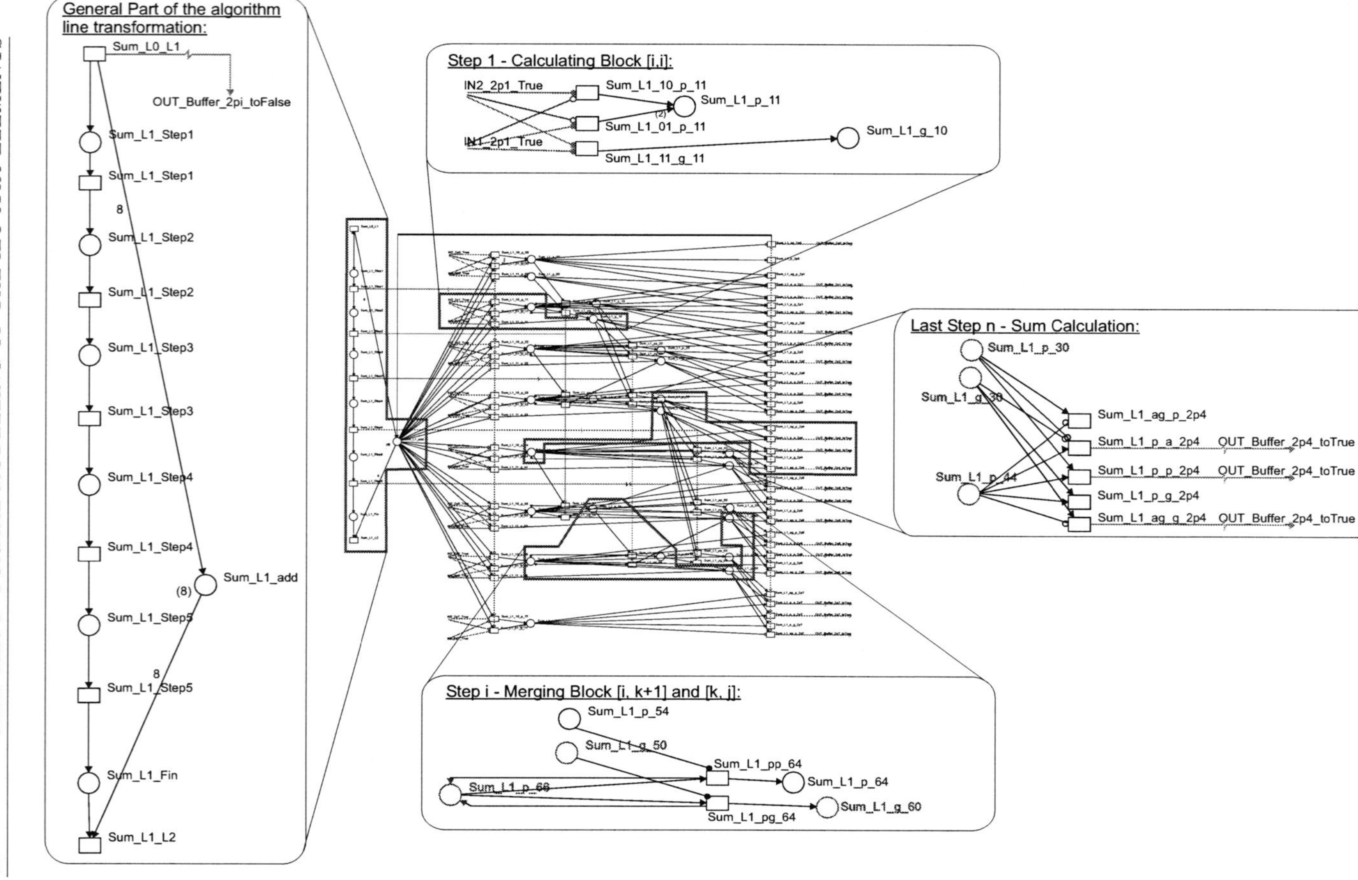

Figure 4.9 NCE structure of a Carry-Lookahead-Adder

4.2.3 Subtraction

Using the derived NCE structure of the base operation addition, it is easy to receive an NCE structure for subtracting variable *IN2* from *IN1*. The reason for this is, that the negation of the integer-valued data *IN2* can be done by connecting the condition and inhibitor arcs of the first evaluating step instead to the place *IN2_ 2pi_ True* to *IN2_ 2pi_ False* and to do the sum calculation with an input carry c_{-1} equal to *true*. This will change the formula of OUT_i to $OUT_i = p_{i,i} \oplus (g_{i-1,0} \vee p_{i-1,0})$ and the truth table to Table 4.2.

OUT_i	$p_{i,i}$		$g_{i-1,0}$	$p_{i-1,0}$	
0	0		0	0	~~not modelled~~
1	0	—o	0	1	$p_{i-1,0} \rightarrow$
1	0	—o	1	0	$g_{i-1,0} \rightarrow$
1	0		1	1	~~not possible~~
1	1	$\rightarrow$	0	0	$p_{i-1,0}$ $g_{i-1,0}$ —o
0	1	$\rightarrow$	0	1	$p_{i-1,0} \rightarrow$
0	1	$\rightarrow$	1	0	$g_{i-1,0} \rightarrow$
0	1		1	1	~~not possible~~

Table 4.2 Truth table of the Carry-Lookahead-Adder (Subtraction)

The first row has not to be modelled because the token remains at the place †_ *add*, and the sum bit OUT_i is switched to *false* at the beginning of the calculation. Furthermore, it will not be possible for a block *[i,0]* to be generating and propagating at the same time (row 4 and 8). As could be seen in Figure 4.10, the number of used transitions is equal to the adder model, only the destination of the event arcs is different, due to the value of OUT_i at the first column of Table 4.2. The used names are also the same.
At the formal model of the servo control system presented in Figure 3.12 at Page 39 *IN1* would get the value of the decoded actual position of the shaft and *IN2* has the value of the reference position. The result *OUT* of the subtraction is the position error.

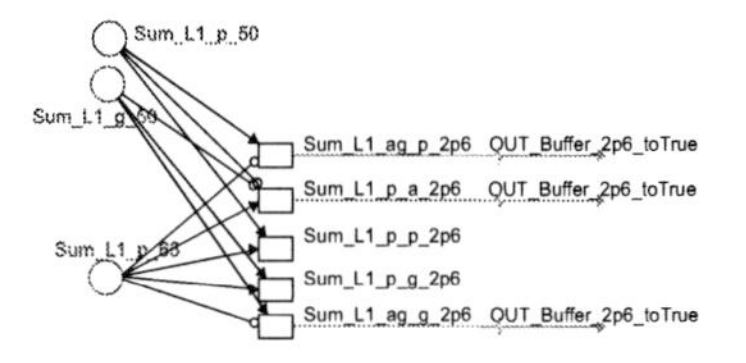

Figure 4.10 Changes to the NCE structure of the Carry-Lookahead-Adder at the subtraction

4.2.4 Increment and Decrement

Special operations on integer-valued data are the increment and decrement to add or subtract a value of 1. Using a Carry-Lookahead-Adder or a Conditional-Sum-Adder for this operation would be as to break a fly on the wheel. As known from the methods of informatics an integer value is incremented by changing bit_0 and every bit_i changes if bit_{i-1} changes from 1 to 0. Otherwise, the operation decrement is realised by changing bit_0 and changing bit_i only if bit_i has changed from 0 to 1.
Furthermore, these operations are not only used at the Master-Controller of the Parametrized Master-Task-Controller approach, but also at the event handling function blocks *E_CTD* (Down counter), *E_CTUD* (Up/Down counter), *E_CTU* (Up counter), *E_TABLE* and *E_TRAIN* specified at appendix A of the [IEC-61499-1]. Thus, these operations may be used quite often and using a special NCE structure to increment and decrement a binary coded integer value, instead of the NCE structure of a Carry-Lookahead-Adder would reduce the number of used places and transitions at the controller model. Further, the number of fired steps would be reduced from $2 + log_2(n)$ to only 1, whatever the number n of used bits is.
Figure 4.11 presents the NCE structure created during the transformation of the first line of the algorithm *CU* of the function block *E_CTU*, which is abbreviated by a † in the following. The transition reaching this line has a post-arc to the place †*_bits* with arc weight of n (number of used bits) as well as a post-arc to the place †*_Increment*. The place †*_bits* has the capacity of *2*n*, because it lies at the pre- and post-set of the transitions †*_biti*. At the post-set of place †*_Increment* lies a transition with the same name and having an outgoing event arc to the transition †*_bit0*. Every transition †*_biti* has an outgoing event arc to the transitions *CV_2pi_Buffer_toTrue* and *CV_2pi_Buffer_toFalse* as well as an incoming event arc from the transition *CV_2pi-1_Buffer_toTrue*. The transition leaving this algorithm line has to remove all tokens from the presented NCE structure. Thus, it has a pre-arc with arc weight of n to the place †*_bits* as well as a pre-arc to the place †*_Fin*.

Comparing the NCE structures of the incrementing and decrementing operations with 8 transitions to the NCE structure of an 8 bit Carry-Lookahead-Adder with 24 places and 78 transitions or an 8 bit Conditional-Sum-Adder with 50 places and 115 transitions it emphasises the necessity of them as well as the already mentioned lower number of fired steps.

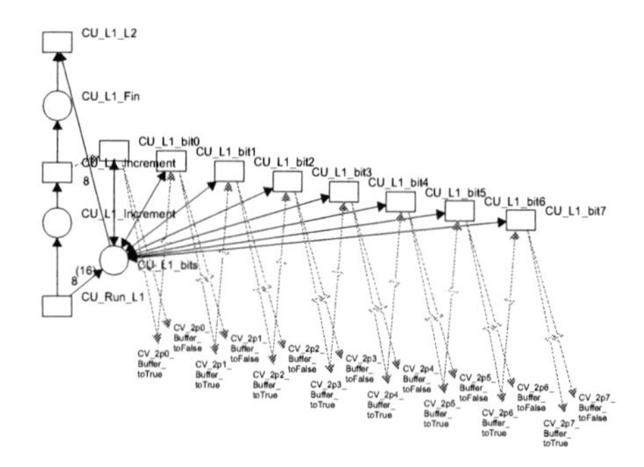

Figure 4.11 NCE structure of the incrementing operation

4.2.5 Comparison

To map the comparison to an NCE structure it has to be checked if it is done between Boolean or integer data values and as well as if it is done between a variable and a static value or between two variables. A boolean comparison, e.g., if the value of a variable is *true* or *false* or if two variables are equal, can simply be modelled by condition and inhibitor arcs. The comparison of an integer-valued data with a static integer value can be modelled as presented in the execution model at the modelling of the condition of the EC transitions. In the following, the comparison between two integer-valued variables should be introduced. As done at the explanation of the Carry-Lookahead-Adder the symbol †, should be used to abbreviate the algorithm name and the line number.
According to the previous description of transforming function blocks to formal models, both variables are available as binary values. Thus, the methods of informatics are applicable and the solution will be a multi-stage one. At first, a decision is taken for every bit i according to the truth Table 4.3, if $IN1_i$ is greater or lower than $IN2_i$.

$IN1_i$	$IN2_i$	$IN1_i = IN2_i$	$IN1_i < IN2_i$	$IN1_i > IN2_i$
0	0	*true*	*undefined*	*undefined*
0	1	*false*	*true*	*false*
1	0	*false*	*false*	*true*
1	1	*true*	*undefined*	*undefined*

Table 4.3 Truth table of the comparison of two bits

Next, it has to be checked if the most significant decision is *greater* or *lower*. This will switch the result of the hole comparison to *greater* or *lower*. If both variables are equal, no decision can be made, and it is *undefined* for every bit i. Figure 4.12 shows the NCE structure of a *greater-than-comparator* for two integer-valued variables *IN1* and *IN2*. On the left, the algorithm, which compares the two variables, is shown.
By entering the comparing model, n tokens are added to the place $\dagger\_undefined$, because there have to be n bits checked. The next fired transition $\dagger\_CompareBits$ is the event source of the transitions $\dagger\_2pi\_toGreater$ and $\dagger\_2pi\_toLower$ to distinguish if variable *IN1* is greater or lower than variable *IN2* at bit i. The modelling of greater and lower is

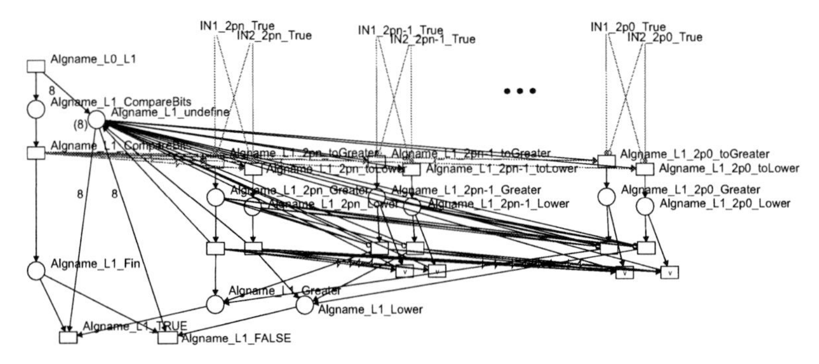

Figure 4.12 NCE structure of a Greater-than-comparator

done according to Table 4.3 by inhibitor and condition arcs. Every fired transition takes one token from the place $\dagger\_undefined$, and only if the value of both variables is equal all tokens remain there. Afterwards, the most significant decision about greater and lower turns the result of the comparator to greater or lower. The modelling is done by a transition at the post of the place $\dagger\_2pi\_Greater$ and $\dagger\_2pi\_Lower$ and connecting them by inhibitor arcs with all places $\dagger\_2pj\_Greater$ and $\dagger\_2pj\_Lower$ and $j > i$. Firing one of these transitions removes the token from the place at the pre-set and stores one token at the place $\dagger\_Greater$ or $\dagger\_Lower$ and again one at $\dagger\_undefined$. Furtherone, the token from all places $\dagger\_2pk\_Greater$ or $\dagger\_2pk\_Lower$ and $k < i$ is removed and transferred back to the place $\dagger\_undefined$, by forcing the connected event sinks with the event mode $\boxed{\vee}$. Thus, at the end place $\dagger\_Fin$ is marked, n tokens are at the place $\dagger\_undefined$ again, and depending on the value of the variables *IN1* and *IN2*, either the place $\dagger\_Greater$ or $\dagger\_Lower$ is marked.
Depending on the transformed algorithm, the result can be used as needed, but it has to make sure to remove all tokens from the given NCE structure as presented with the transitions $+\_$ *TRUE* and $+\_$ *FALSE*.

4.2.6 Derived Modelling Rules

Before the presented examples and the derived transformation rules could be used to model the data processing of IEC 61499 function blocks, a lexical analysis has to be done for each algorithm by a priori defined formal grammar. This would incorporate the parsing of the syntax and a semantic analysis to identify each statement and the used variables. In this work, this is done for a limited set of *Structured Text*, because this is the implementation language used at any algorithm of the function blocks presented in Chapter 3. Nonetheless, it is possible to use all [IEC-61131-3] programming languages as *Instruction List (IL), Ladder Logic (LD), Function Block Diagrams (FBD)* and *Sequential Function Chart (SFC)* as well as high level programming languages like *Java, C++* and *Delphi.* Thus, a syntax parser in combination with a lexical analysis as well as a semantic checker have to be implemented for all of them in the near future to improve the function block transformation. A good starting point would be the *MatIEC* compiler developed at a collaborative project between the *University of Porto* and the *TBI SARL - Lolitech* and being part of the open source framework Beremiz[1], which is described in more detail at [TBdS07]. The *MatIEC* compiler is normally used to translate a whole PLC configuration, but it can be used with function blocks conforming to the [IEC-61131-3] as well. Then it performs a lexical analysis and parses the syntax in one step and creates an abstract syntax tree. Next, this syntax tree is checked for semantic correctness and generates in step 4 equivalent C code. As described in [GPH10] this last step 4 has to be replaced by the generation of NCE structures according to the following rules. These generated NCE structures can be imported at the function block transformation for each algorithm. The benefit of this approach is, only the generation has to be implemented once, and all the rest is already done for *IL, ST* and *SFC* and therefore also for *LD* and *FBD.* But as previously mentioned in this work it was enough to do the implementation for a limited set of ST.
These subrules presented in the following will extend transformation rule 4 of the execution model. Thus, the general part of this rule remains untouched to provide the connection between the execution and data processing model. At the conditional part, it has to be distinguished between Boolean and integer-valued data, which leads to the following rule.

[1] http://www.beremiz.org

Transformation rule 4.1 - Conditional Part: Each line i of the algorithm is activated by a transition named *AlgName_ Li-1_ Li* and finished by a place named *AlgName_ Li_ Fin*. Further, the name of each transition and place gets a prefix consisting of the algorithm name and the line number *AlgName_ Ln* (abbr. †), where n stands for the current line number. First, it has to be checked if Boolean or integer-valued data is used at the algorithm line by checking for each variable if there exists a place named +_ *True*, where + is the name of the variable. If it is true, only boolean data is used, and this line has to be transformed according to rule 4.1.1 and otherwise by 4.1.2.
Both rules have subrules to set the variables of this type to certain static values, to perform a value assignment by given equations and to compare them.

Transformation rule 4.1.1 - Boolean Data:

Transformation rule 4.1.1.1 - Reset - Set Boolean Data: If the boolean equation only sets or resets a boolean data output, an event arc has to connect the transition activating this line with the transition +_ *Buffer_ toFalse* to reset it or to the transition +_ *Buffer_ toTrue* to set it. Thereby, the symbol + stands for the name of the data output or the internal variable.
Figure 4.13 shows the transformation result of an algorithm setting and resetting a Boolean data output.

Transformation rule 4.1.1.2 - Boolean equation: At boolean equations, the law of distribution and de Morgan have to be used to resolve all brackets at the term after the equation sign to get n conjunctive terms connected by disjunctions. Each of them is modelled by a transition with the firing mode *instantaneous* and is named †_ j, where j stands for the number of the conjunctive term, and every transition is connected by an event arc to the transition +_ *Buffer_ toTrue*. Furthermore, there is a transition with the firing mode *spontaneous* connected by an event arc to the transition +_ *Buffer_ toFalse*, which will only

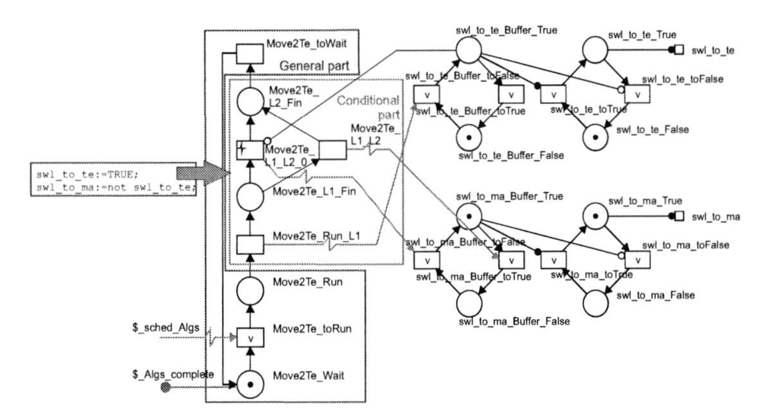

Figure 4.13 Transformation result of an algorithm with Boolean data

fire if none of the other transitions is condition-enabled. All these transitions have at the pre-set the final place of the algorithm line before and at post-set the final place of this algorithm line.

At the second line of the transformed algorithm in Figure 4.13 the Boolean equation $swl\_to\_ma = \neg swl\_to\_te$ is transformed.

Transformation rule 4.1.1.3 - Boolean comparison: A boolean comparison checks if the value of two terms is equal or not. Similar to the transformation of boolean equations, all brackets at both terms have to be resolved first by the law of distribution and de Morgan to receive n conjunctive terms connected by disjunctions.

If the left and right term consist only of a single variable, the transformation of an

equal to comparison is done by inserting two transitions with the firing mode *instantaneous* and connecting one of them by condition and the other by inhibitor arcs to the places $+\_$*True*, where $+$ stands for both variable names.

not equal to comparison of these two variables is done by inserting also two transitions with the firing mode *instantaneous* and connecting the first with an inhibitor arc to the place $+\_$*True* of one variable and by a condition arc to the place $+\_$*True* of the other variable. The second transition is connected vice versa by signal arcs.

Both inserted transitions have at the pre-set the final place of the algorithm line before and at post-set the place $\dagger\_True$. Furthermore, a transition with the firing mode *spontaneous* connecting the place of the algorithm line before and $\dagger\_False$ has to be inserted.

Applying the transformation rule to the comparison $a = b$ and $a \neq b$ the transformation result of algorithm includes the NCE structures in Figure 4.14a and 4.14b.

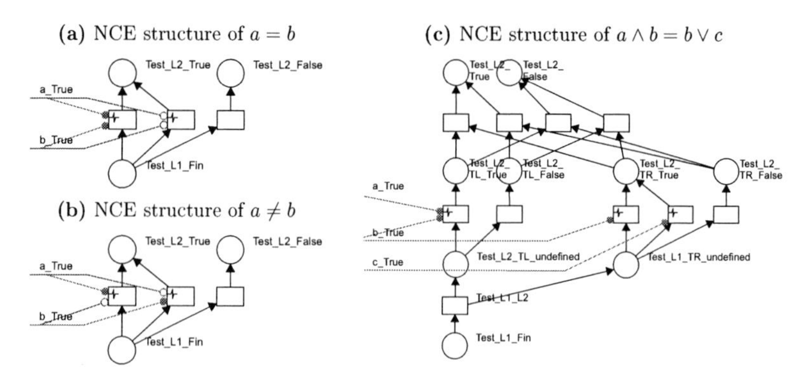

(a) NCE structure of $a = b$

(b) NCE structure of $a \neq b$

(c) NCE structure of $a \wedge b = b \vee c$

Figure 4.14 Transformation of a Boolean comparison

If the left and right term consist of n conjunctive terms, the transformation is separated into the steps, calculation of the value of each term (1) and comparing the values (2).

1. Two places †_ *TL_ undefined* and †_ *TR_ undefined* have to be inserted. These places are connected by the transition *AlgName_ Li-1_ Li* to the final place of the previous algorithm line. The transformation of the left n conjunctive terms results in n *instantaneous* and one *spontaneous* transition. The *instantaneous* transitions are connected by condition and inhibitor arcs to the places +_ *True*, according to the represented conjunctive term, where + stands for the variable name. At the pre-set of each transition, place †_ *TL_ undefined* is located, and at the post-set the place †_ *TL_ True*. The *spontaneous* transition has no incoming signal arc and connects the places †_ *TL_ undefined* and †_ *TL_ False*. The transformation of the right n conjunctive terms is done similarly.

2. If an

   equal to comparison is transformed a transition with the places †_ *TL_ True* and †_ *TR_ True* and a transition with the places †_ *TL_ False* and †_ *TR_ False* at the pre-set have to be inserted and connected at the post-set to the place †_ *True*.
   Further, a transition connected to the places †_ *TL_ True* and †_ *TR_ False* and a transition connected to the places †_ *TL_ False* and †_ *TR_ True* have to be inserted. Both have at the post-set the place †_ *False*.

   not equal to comparison is transformed the post-arcs of four transitions have to be exchanged.

Transformation rule 4.1.2 - Integer-valued Data:

Transformation rule 4.1.2.1 - Set Integer-valued Data: To set a variable with an integer-valued data type to

a static value *n* , event arcs have to connect the transition activating this algorithm line with the transition +_ *2pi_ Buffer_ toFalse* or +_ *2pi _Buffer_ toTrue* according to the binary representation of the value n.

to the value of another variable *Y* , two transitions have to be inserted for each bit i named †_ *bit0_ toTrue* and †_ *bit0_ toFalse*. The first transition is connected by a condition arc and the other by an inhibitor arc to the place *Y_ 2pi_ True*. Additionally, the first transition has as event sink +_ *2pi_ Buffer_ toTrue* and the other has +_ *2pi_ Buffer_ toFalse* as event sink. Both transitions have as event source the transitions inserted at $i-1$ and the event mode $\boxed{\vee}$ and the place †_bits at the pre- and post-set. This place has the capacity of $2*n$, where n is the number of used bits. Further, this place is connected by a pre-arc to the transition reaching this line and by a post-arc to the transition leaving this line. Both arcs have the weight of n.
Finally, the places †_ *set* and †_ *Fin* are inserted and connected by the transition †_ *set*, which has an outgoing event arc to the transitions †_ *bit0_ toTrue* and †_ *bit0 _toFalse*. The place †_ *set*is located at the post-set of the transition reaching this line, and the place †_ *Fin* is part of the pre-set of the transition leaving this line.

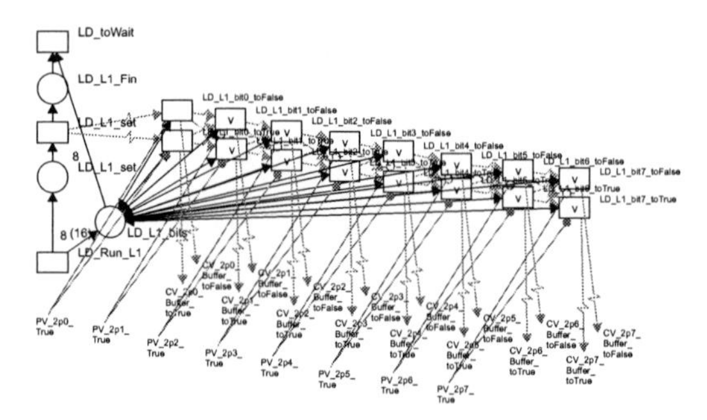

Figure 4.15 NCE structure to set an integer-valued variable CV to the value PV

During the transformation of the function block *E_CTD* (Counter Up) defined at the Appendix A of the standard, the algorithm LD has to be transformed, which sets the value of the data output CV to the value of the data input PV. Thereby, transformation rule 4.1.2.1 was used and the NCE structure of Figure 4.15 created.

To define a transformation rule for adding the values of two data inputs *IN1* and *IN2* and store the result at the buffer of the data output OUT, the NCE structure representing the Carry-Lookahead-Adder from Section 4.2.2 should be taken. The reason is, that this NCE structure needs less steps to calculate the new value of the data output OUT, and despite the NCE structure of a Conditional-Sum-Adder it takes less time to calculate the steps. The reason therefore is that the Carry-Lookahead-Adder evaluates only the attribute *propagate* and *generate* and the Conditional-Sum-Adder calculates the *sum* and *sum+1* as well as the *carry* and *carry+1* for each block [i,i] and afterwards the blocks are merged [LF80, CC06].

Transformation rule 4.1.2.2 - Add Integer-valued Data: For transforming the addition of two *n-bit* integer-valued variables *IN1* and *IN2* and as result OUT, there have to be inserted first a sequence of $2 + log_2(n)$ places and transitions named †_*Stepi*. The place †_*Step1* is located at the post-set of the transition activating this algorithm line, and the transition †_*Stepj* is at the pre-set of the place †_*Fin* ($j = 2 + log_2(n)$). Next it has to be inserted the place †_*add* having a pre-arc with the arc weight n to the transition activating this algorithm line and a post-arc with the same arc weight to the transition reaching the next algorithm line. Further, the transition activating this line has event arcs to every transition *OUT_Buffer_2pi_toFalse*.
Now the model of the Carry-Lookahead-Adder has to be inserted step by step as described in Section 4.2.2.

1. For every bit i 3 transitions named †_*10_p_ii*, †_*01_p_ii* and †_*11_g_i0* with the place †_*add* at the pre-set and an incoming event arc from the transition †_*Step*1 have to be inserted. The first two transitions are connected to the place †_*p_ii*, which has a capacity of two, by a post-arc. The third transition has a post-arc to the place †_*g_i0*. The first transition has a condition arc to the place *IN2_2pi_True* and an inhibitor arc to the place *IN1_2pi_True*, and the second transition vice versa. The third transition has a condition arcs to both places.
   Inserting of the transition †_*11_g_i0* and the place †_*g_i0* and all connecting arcs is neglected at the most significant bit.

At the least significant bit the pre-arcs of place †_*p*_*ii* have the arc weight 2.

2. For every bit $i = 2 * 2^x * y + 2^x + z$ with

$$i < n - 1 \wedge x < log_2(n) \wedge y < \frac{n}{2^{x+1}} \wedge z < 2^x$$

2 transitions named †_*pp*_*ij* and †_*pg*_*ij* with the event source †_*Stepx* as well as a place named †_*p*_*ij* have to be inserted ($j = 2 * 2^x * y$).
At the pre-set of the both transitions is the place †_*p*_*ik*. If $i = k$ then this place is also at the post-set of the transitions ($k = 2 * 2^x * y + 2^x$).
Further, the transition †_*pp*_*ij* has a post-arc to the place †_*p*_*ij* and a condition arc to the place †_*p*_*lj*. The transition †_*pg*_*ij* has a post-arc to place †_*p*_*i0* already inserted at step 1 and a condition arc from place †_*g*_*lj* ($l = k - 1$).
At the most significant bit this is neglected.

3. For every bit i five transitions named †_*ag*_*p*_*2pi*, †_*p*_*a*_*2pi*, †_*p*_*p*_*2pi*, †_*p*_*g*_*2pi* and †_*ag*_*g*_*2pi* with a post-arc to the place †_*add* and an event arc from the transition †_*Stepj* have to be inserted ($j = 2 + log_2(n)$). Every transition with the prefix †_*p* has a pre-arc to the place †_*p*_*ii* and all other an inhibitor arc to this place. The transitions †_*p*_*a*_*2pi*, †_*p*_*p*_*2pi* and †_*ag*_*g*_*2pi* have an event arc to the transition named *OUT*_*Buffer*_*2pi*_*toTrue*.
Further, the transitions with the prefix †_*ag*_*p* and †_*p*_*p* have a pre-arc to the place †_*pj0*, the transitions †_*p*_*g* and †_*ag*_*g* have a pre-arc to the place †_*gj0*, and the transition †_*p*_*a* has two inhibitor arcs to the places †_*pj0* and †_*gj0* ($j = i - 1$).

Transformation rule 4.1.2.3 - Subtract Integer-valued Data To transform the subtraction of one *n-bit* integer-valued variable *IN2* from *IN1* and storing the result at *OUT*, points 1 and 3 of the previous adding rule have to be changed. At 1 the destination of the condition and inhibitor arcs have to be changed from *IN2*_*2pi*_*True* to *IN2*_*2pi*_*False*. At 3 the event arcs to the transition *OUT*_*Buffer*_*2pi*_*toTrue* have to be changed as follows:

- †_*ag*_*p*_*2pi* new event arc
- †_*p*_*p*_*2pi* no event arc

To develop a transformation rule to compare to integer-valued variables *IN1* and *IN2* the presented NCE structure in Section 4.2.5 should be taken.

Transformation rule 4.1.2.4 - Compare Integer-valued Data The transformation of the comparison of two *n-bit* integer-valued variables *IN1* and *IN2* to the formal model is done by inserting the places †_*CompareBits* and †_*Fin* and the connecting transition †_*CompareBits*. Further, the place †_*undefined* with the capacity of n is inserted. At the pre-set of this place the transition activating this algorithm line is located. The pre-arc has the arc weight of n.
Now, the model of the comparator as described in Section 4.2.5 starting with the creation of the places †_*Greater* and †_*Lower* has to be inserted.

1. Two transitions named †_*2pi*_*toGreater* and †_*2pi*_*toLower* as well as two places †_*2pi*_*Greater* and †_*2pi*_*Lower* are inserted for every bit i. At the pre-set of

every transition the place †_*undefined* is located and additionally every transition has an incoming event arc from transition †_*CompareBits*. Further, transition †_*2pi_toGreater* has a condition arc to the place *IN1_2pi_True* and an inhibitor arc to *IN2_2pi_True*. The transition †_*2pi_toLower* is connected vice versa by the signal arcs.

2. For every bit i two transitions named †_*2pi_Greater* and †_*2pi_Lower* are inserted and connecting the places †_*2pi_Greater* and †_*Greater* as well as connecting the places †_*2pi_Lower* and †_*Lower*. Every transition has an incoming inhibitor arc from all places †_*2pj_Greater* and †_*2pj_Lower* $(j > i)$.
Except at the most significant bit, two transition are inserted with the event mode $\boxed{\vee}$ having the place †_*2pi_Greater* or †_*2pi_Lower* at the pre-set and an incoming event arc from the transitions †_*2pj_Greater* and †_*2pj_Lower* $(j > i)$.
All transitions have at the post-set the place †_*undefined*.

4.3 Scheduling Model

As mentioned in Section 1.1 the verbose description for the function block execution and the scheduling of function blocks inside function block networks leads to several runtime implementation complying with the standard, but using different semantics. This may lead to different plant behaviours as explained in Section 3.5. At the presented example, the function block network controlling the Gripper Station is moved from the FBRT to FORTE runtime, and the plant behaviour changed from the desired behaviour to executing each time the same action. To predict such situations before running the plant, the formal model used for verification should be extended at this section by a scheduling model. This should incorporate every implemented scheduling scenario of function blocks inside a function block network. Thus, it should be possible for any control engineer to verify the implemented control application for every or at least the used runtime. According to the standard, the scheduling function is part of the resource of a device. Thus, resource models will be defined at this section.

4.3.1 Single- or Multi-threaded resource

The NCE module in Figure 4.16a represents the execution model of a resource with one thread and no synchronous execution behaviour. It consists of the different places *Idle*, *Executing_FB* and *Enable_EvOutputs*, and the first and last place have a condition output arc. The condition output *Idle* of the presented module will be connected to the condition input *Resource_Idle* of all NCE modules representing a function block mapped to this resource. Thus, inside the NCE structure created during transforming an event input, the transition *_*Release* is condition-enabled if the place *Idle* of the resource module is marked. If this transition forces also a transition modelling an EC transition, the NCE module representing a function blocks issues the event *FB_scheduled*, which is connected to the event input of the resource module (Figure 4.16). Thus, if a modelled EC transition fires, the token will move from place *Idle* to *Executing_FB*. Now, the modelled algorithms of the scheduled function block are executed, and if the place *WAIT* is reached again, the modelled EC state gets into the *$_Idle* state, and therefore the transformed function block

(a) Single- /Multi-threaded resource (b) Resource with Synchronous Execution (c) Resource with Cyclic Execution

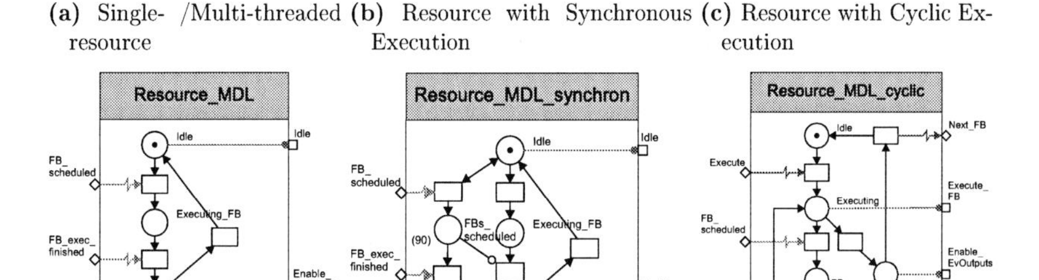

Figure 4.16 NCE modules to model resources with different scheduling functions

into *FB_ idle*. Thereby, the event *FB_ exec_finished* is issued and forces the module of the resource to switch into *Enable_ EvOutputs*. This enables the transition *_ *Release* of all NCE structures representing an event output of the function blocks mapped to this resource. Due to the fact that these transitions have the firing mode instantaneous, all of them will fire, before the resource model gets back to the state *Idle*, which will enable the event input transitions again. Due to the semantic of the formal model, all enabled EC transitions fire concurrently, and therefore all transformed function blocks are scheduled concurrently with this resource model.
If the resource uses more than one thread with a static function block assignment as described in [TD06a], the above described module has to be used as often as threads are available, and the NCE modules of the assigned function blocks have to be connected to the NCE module describing the thread behaviour. Another approach discussed in [TD06a] is a multi-threaded resource without a static assignment of function block instances. This can be modelled by increasing the number of tokens at the place *Idle*. Thus, as long as there are threads or respectively tokens available, the function blocks are scheduled.

4.3.2 Resource with Synchronous Execution

If the used resource performs instead of a single-threaded or a parallel multi-threaded execution a truly synchronous execution of function blocks, the used NCE module describing the resource execution has to be replaced by Figure 4.16b. The model incorporates the same states, but the transition between the place *Idle* and *Executing_ FB* is no longer forced by an input event. Thus, the transition *_ *Release* of all NCE structures representing an event input of function blocks mapped to this resource fire before the token moves from *Idle* to *Executing_ FB*, because these transitions have the firing mode instantaneous. If thereby a modelled EC transition fires and forces the modelled function block into the state *FB_ scheduled*, the resource module receives the input event *FB_ scheduled* and forces the transition situated at the pre- and post-set of the place *idle* and the place *FBs_ scheduled*. Thus, the number of tokens at the place *FBs_ scheduled* represents the number of synchronously executed function blocks. Only, if all modelled function blocks get into the state *FB_ idle* again and issue the event *FB_ exec_finished*, all tokens are removed from the place *FBs_ scheduled*, and the transition between *Executing_ FB* and *Enable_ EvOutputs* gets condition-enabled. As described above, the transition back to *Idle* has the firing mode spontaneous and fires after all transitions *_ *Release* of all NCE structures representing an event output of mapped function blocks. Thus, two synchronisation points are modelled. The first is the evaluation of EC transitions, and the second is the publishing of output events.

4.3.3 Resource with Cyclic Execution

To model a resource with cyclic execution behaviour, a set of the NCE modules shown in Figure 4.16c has to be used. In addition to every NCE module representing a transformed function block of the resource, such a module is inserted. Depending on the cyclic execution of the function blocks, the additional modules are interconnected by event arcs from *Next_FB* of the predecessor module to *Execute* of the successor module. The last module of the cycle is connected to the first one.
The condition output *Execute_FB* is connected to the condition input *Resource_Idle* of the transformed function block this NCE module is assigned to. Thus, a token at the place *Execute_FB* condition enables the transition **_Release* of all NCE structures representing an event input. If now a transition modelling an EC transition is forced by a transition **_Release*, the model of a connected basic function block issues the event *FB_scheduled.* This forces the resource model into the state *FB_scheduled* and the transition **_Release* of all other connected NCE structures representing an event input are not condition-enabled any longer. Only, if the model of the basic function block issues the event *FB_exec_finished* these transtion get condition-enabled again. Due to the fact, that all transitions **_Release* have the firing mode instantaneous, all active events of a connected basic function block are released first, before the transition between the places *Executing* and *Enable_EvOutputs* of the resource module fires. Now, all output events are released, before the token at the resource modules flows back to the place *Idle* and trigger the next resource module connected to another basic or composite function block.
The mentioned model of a basic function block may be the model of the function block connected to the resource module itself, or be a model of a component function block of a connected composite function block.

4.3.4 Derived Modelling Rules

Transformation rule 5 - Function Block Network: For each component function block of a composite function block and each function block instance of a resource, the prior transformed NCE module is inserted. Each event connection between component function blocks or function block instances is transformed to an event interconnection between the submodules with the same source and sink. For each data connection, the actual data type has to be checked and if it is Boolean one rule 5.1 has to be applied. Otherwise, rule 5.2 or 5.3 depending on the used data type is applied.
Afterwards, it has to be distinguished if the function block network is part of a resource of part or a composite function block.

At the transformation of a

composite function block *X* each event connection with the source at the event input * is transformed to an event interconnection between the event output * of the NCE module *X_ inputs* and the original sink. Also, if an event connection from a component function block to an event output * should be transformed, an event interconnection between the original source and the sink *X_ outputs.** is inserted.
According to the number n of component function blocks inside the function block network, two NCE modules of the type *E_ Merge_ n* have to be inserted to merge output events *FB_ scheduled* and *FB_ exec_ finished* of all sub modules. The output event *EO* of the module merging the events *FB_ scheduled* is connected to the event output *FB_ scheduled*, and the output event *EO* of other event merging NCE module is connected to the event output *FB_ exec_ finished.*

Figure 4.17 shows the transformation result for the composite function block to control the Gripper Station of Figure 3.11. During the transformation of the interface of the composite function block, the NCE modules *Gripper_ Reuse_ inputs* and **_ outputs* are created to represent the interface and to realise the sampling of the data inputs and the publishing of the data outputs, as already presented in Figure 4.1b. To ease the recognition of the transformed function block network, the same colour scheme is applied to the NCE modules representing a transformed function block. All white coloured NCE modules are additional modules to merge the events *FB_ scheduled* and *FB_ exec_ finished* as well as to model the inputs and outputs of the transformed composite function block. All transformed event connections of the function block network are red coloured event interconnections between the submodules, and all others are parts of the scheduling model. Each transformed data connection is represented by n blue coloured condition interconnections between the submodules. All black coloured one are parts of the scheduling model and connect the condition inputs *Resource_ Idle* and *Enable_ EvOutputs* with the equal named condition inputs of the submodules.

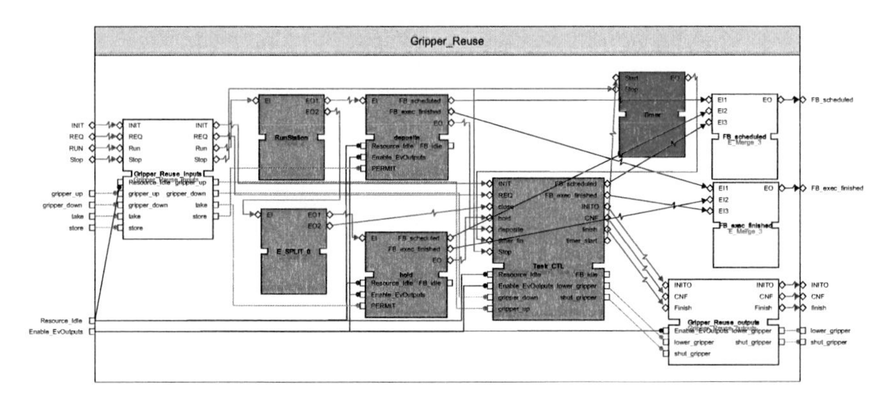

Figure 4.17 NCE module of the composite FB Gripper_Reuse

At the transformation of a

resource one or more of the presented resource modules are inserted to model the scheduling of the function block instances.

If the resource has a *single-threaded* or *synchronous* execution behaviour, then one of the corresponding modules is inserted. Afterwards, the condition outputs *Idle* and *Enable_EvOutputs* of this module are interconnected to condition inputs *Resource_Idle* and *Enable_EvOuputs* of the NCE modules representing a transformed function block instances. According to the number n of function block instances being part of the resource owned function block network, two NCE modules of the type *E_Merge_n* have to be inserted to merge the output events *FB_scheduled* and *FB_exec_finished* of all submodules. The output event *EO* of the module merging the events *FB_scheduled* is connected to the event input *FB_scheduled* of NCE module representing the scheduling function of the resource and the output event *EO* of other event merging NCE module is connected to the event input *FB_exec_finished.*

If the resource has a *multi-threaded* execution behaviour with a dynamic assignment of function blocks to a thread, then only the number of tokens at the place *Idle* of the NCE module shown in Figure 4.16a has to be increased up to the number of available concurrently running threads.

If the resource has a *multi-threaded* execution behaviour with a static assignment of function blocks to a thread, then for each thread, an NCE module as presented in Figure 4.16a has to be inserted. Afterwards, the condition outputs *Idle* and *Enable_EvOutputs* of the NCE module modelling a thread are connected to the condition inputs *Resource_Idle* and *Enable_EvOuputs* of the NCE modules representing the transformed function block instances assigned to this thread. According to the number n of function block instances assigned to this thread, two NCE modules of the type *E_Merge_n* have to be inserted to the merge output events *FB_scheduled* and *FB_exec_finished* of all submodules representing a transformed function block instance assigned to this thread.

If the resource uses a *cyclic* execution semantic of function block instances, then n NCE modules of the type *Resource_MDL_cyclic* have to be inserted according to the number of instantiated function blocks. Each of these modules is connected by an event interconnection from the event output *Next_FB* to the event input *Execute* of the successor modules. The last module of this chain is connected to the first one. The condition output *Execute_FB* is connected to the condition input *Resource_Idle*, and *Enable_EvOutputs* is connected to the input *Enabled_EvOutputs* of NCE module representing the function block instance being cyclical activated. The event outputs *FB_scheduled* and *FB_exec_finished* of the prior connected NCE module are connected to the correspondent event inputs of the resource module activating this module.

Figure 4.18 presents the formal model of the servo control device with one resource having a single-threaded execution behaviour. As could be seen, the function block network of

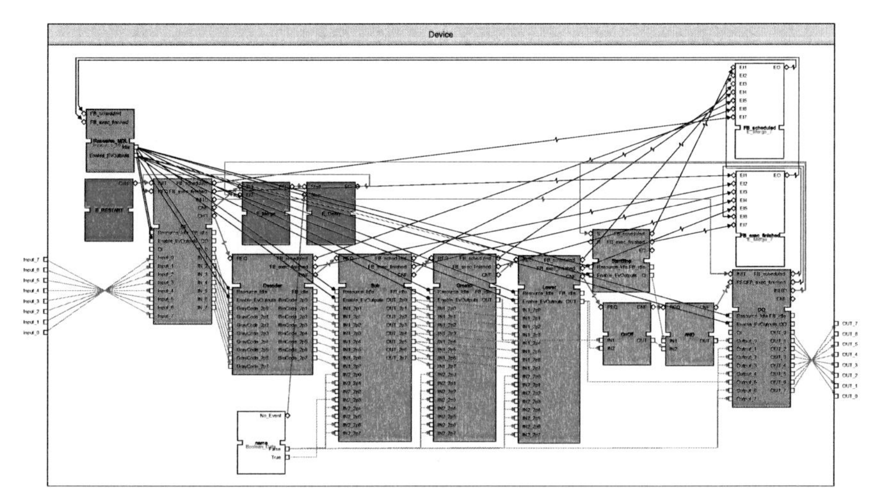

Figure 4.18 NCE module of the Servo-Control device with a Single-threaded resource

Figure 3.12b is extended by the function block *E_Restart* during the mapping to a resource to initiate the start event of the resource. As described at the transformation of the function block network inside a resource, all event interconnections resulting from a prior event connection between function block instances are red coloured, and all condition interconnections resulting from data connection are blue coloured. The white coloured NCE modules at the right are inserted to merge the event outputs *FB_scheduled* and *FB_exec_finished*. At the top left, the NCE module modelling the scheduling function of a single-threaded resource can be found. As the last step, the condition inputs of the module representing a model of the SIFB, explained in detail in Section 5.2, to read the inputs are interconnected to the correspondent condition inputs of the NCE module describing the device. Also the condition outputs of the module describing the SIFB to write outputs are interconnected to condition outputs of the device module.

Transformation rule 5.1 - Boolean Data Connection: Each Boolean data connection between component function blocks of function block instances is transformed to a condition interconnection between the submodules with the same source and sink. At the transformation of a composite function block *X*, each data connection from the data input + is transformed to a condition interconnection between *X_inputs.+* and the original sink. Also, if a data connection of the composite function block *X* to a data output + should be transformed, a condition interconnection between the original source and the sink *X_outputs.+* is inserted.

Transformation rule 5.2 - Integer-valued Data Connection: For each integer-valued data connection the number n of representing bits have to be retrieved first. Afterwards, transformation rule 5.1 is applied n times with the iterator i and providing the string *+_2pi* for the source and the sink.

Transformation rule 5.3 - Data Connection with an array: For each data connection with an array as data type, the elementary data type of the array has to be retrieved first. Either the transformation rule 5.1 or 5.2 is applied next, depending on the elementary data type and providing the string *+_i* for the source and sink.

4.4 Summary

Modelling as many details as necessary, an Execution Model is developed for different function block types at the beginning of this chapter. It includes the transformation of the interface with sampling and publishing the data, only if an associated event occurs and incorporates Boolean as well as integer-valued data in- and outputs. But also arrays as data type are possible. The Execution Control Chart and a general part of algorithms of basic function blocks are also part of the Execution Model. The transformation of the function block network inside a composite function block is part of the later defined scheduling model.
Thereafter, a Data Processing Model is defined to handle not only Boolean data processing, but integer-valued operations as well. This model is derived from already proven methods of informatics and includes NCE structures to add, subtract, increment, decrement and compare in a binary way the integer-valued data.
Finally, a Scheduling Model is derived to model the different possible scheduling functions a resource complying with the standard can have, to schedule the function blocks inside. This Scheduling Model incorporates as well the transformation of the function block network with the event and data connections.
For each model several rules were defined by the author and used to implement another *SWI-Prolog* module for the *TNCE-Workbench* to automatically transform the function blocks as shown in the following chapter.

Chapter 5

Closed-Loop Modelling

The modelling of the closed-loop systems starts with the automatic transformation of the used function blocks into NCE modules. This automatic transformation is implemented as an additional SWI-Prolog module of the already mentioned TNCES-Workbench (see Section 2.4). After the transformation of all used basic and composite function blocks, the $_D$TNCE module of the device has to be created by incorporating an model of each resource as well as an appropriated model of the process and communication interface realised by service interface function blocks (SIFB). Since the process interface is used to control the process, and the communication interface is used to exchange data with other devices, the condition in- and outputs of the SIFB have to be connected by condition interconnections to the interface of the device module. Thus, the plant and the control model are connected to each other by condition arcs as shown in Figure 5.1. On the one hand, this ensures the absence of event loops from the control device through the plant back to the control. On the other hand, this fits better to the reality than exchanging events between the controller and the environment, because the input and output signals have a steady state for a certain time. Furthermore, how should a controller detect an event from the plant, occurring once at a certain time point? The last step is to model the plant by using the predefined $_D$TNCE modules in Section 2.5 and connecting both received modules to the $_D$TNCE system.

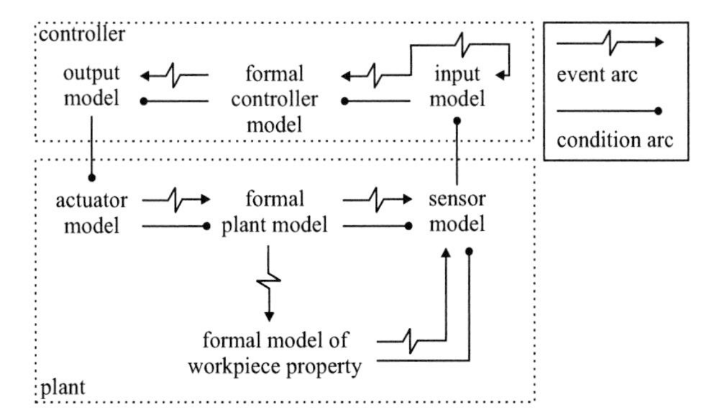

Figure 5.1 Interconnections of an NCES based closed-loop system

5.1 Transformation Process

As introductorily mentioned at the beginning of the chapter, each transformation of a distributed control system into the formal modelling language *discrete timed Net Condition/Event Systems*starts with the transformation of all basic function blocks, if they are not already transformed. Therefore, the TNCES-Workbench provides the menu entry *IEC 61499 - Transfom Basic FB*. After activating the entry, the automatic transformation process starts with choosing the original function block file and building a *Document Object Model* of the provided file. Because of the defined requirements for software tools at the second part of the [IEC-61499-2] standard, it does not matter which engineering environment is used to develop the function blocks. At this thesis mainly the *Function Block Development Kit* (FBDK) as well as the *Framework for Industrial Automation and Control* (4DIAC) [SRE+08a] are used, but despite this also the tools *OOONEIDA - FBench*, *Corfu Engineering Support System* and *nxtStudio* can be used as the source for the transformation process and starting point for the closed-loop modelling of the distributed control system.
After the *Document Object Model* is created, it is passed through to the prolog module *61499-translation* implementing the rule *translate_basic/2* as presented at the Source Code 5.1. At first the dynamic knowledge base is emptied from all dynamic facts, maybe inserted at a previous transformation (line 4-6). Also, the flag to count the number of used transitions and places is set to 1 (l. 7). Thereafter, the parsed *Document Object Model* is transformed into a new one by unifying the rule *element_child/2*(l. 11). This rule is implemented for the root element *FBType* and returns, if proven to be true, a root element representing an NCE module conforming to the markup language presented in Section 2.3 (l. 19).
Next, the interface of the function block is retrieved and stored at the list *Interface* (l. 24). All elements of the function block are stored at the list *Elements* and the position of the most left element is stored at the variable *MinPosX* and the position of the most right element at the variable *MaxPosX* (l. 25). Both positions are later used to determine the position of the interface and the left and right border of the new NCE module. Using the list of elements, a decision can be made, if the actual function block is a simple, a basic or a composite one (l. 30). According to this decision, the rules *create_new_interface/5* and *translate_basic_function/4* are used to transform the interface, the ECC and the algorithms by inserting the presented NCE structures of the previous chapter into the dynamic knowledge base (l. 31-36). Despite this, the rule *create_composite_interface/4* creates the TNCEM files **_inputs* and **_outputs* at the transformation of the interface of a composite function block (l. 38). Both module types are used as module instances and connected according to the function block network transformation rule with the module instance of the prior transformed function blocks (l. 39) during the proving of rule *translate_composite_function/4*.

Source Code 5.1 Function Block Transformation with SWI-Prolog

```
1  translate_basic([],[]):-!.
2  translate_basic(XML,New):-
3          write(XML),nl,
4          (   (   retractall(transition(_,_,_,_,_,_)),  retractall(place(_,_,_,_,_,_)),
5                  retractall(prearc(_,_,_,_)), retractall(postarc(_,_,_,_)), retractall(
                     evarc(_,_,_)), retractall(condarc(_,_,_)), retractall(inhibarc(_,_,_)),
6                  retractall(basicFB(_)),
7                  flag(tnr, T, 1), flag(pnr, P, 1)
8              );
9              true),
10         content(XML,Content),
11         element_child(Content,New),
```

```
12         write(New),nl.
13
14 content(document(_Type, [Content]), Content).
15 content([Content], Content).
16 content(element(A,B,C), element(A,B,C)).
17 content([pi(_XML), Content], Content).  % bit of a hack to handle XML files
18
19 element_child(element('FBType',Attributes,Content),element('FBType',['X'=160, 'Y'=281, '
     Num'=1, 'LocNum'=1, 'Name'=Name, 'Comment'='_', 'Width'=120.0, 'Height'=FBHeight],[
     TNCESInterface,SNS])):-
20         member('Name'=Name,Attributes),
21         format('| Function Block:  ~w ~n',[Name]),
22         %element_child(Content,Children),
23         delete(Content, ' ', ClearContent),
24         get_FB_Interface(ClearContent, Interface),
25         get_used_elements(ClearContent, Elements, MaxXPos, MinXPos),
26         format('MaxPos: ~w ~t MinXPos: ~w~n',[MaxXPos,MinXPos]),
27         XPosR is MaxXPos + 500,
28         XPosL is MinXPos - 200,
29         !,
30         (   memberchk(element('BasicFB',_,_),ClearContent)
31         ->  %transformation of a basic or simple function block
32             %interface
33             create_new_interface(Interface, XPosL, XPosR, TNCESInterface, FBHeight),
34             %translation of the basic
35             translate_basic_function(Elements, XPosL, XPosR, SNS),
36             format('---------- End of Basic Function Block:  ~w~n',Name)
37         ;   %transformation of a composite functoin block (has to be implemented)
38             create_composite_interface(Interface, XPosL, XPosR, TNCESInterface, FBHeight),
39             translate_composite_function(Elements, XPosL, XPosR, SNS)
40         ).
```

Exemplarily for the *transformation rule 2.3 - Event Input*, described at page 46, the implementation of the transformation rules should be explained (Source Code 5.2). Each event input of the function block to be transformed is represented as a tuple consisting of the *Name* and the list *Withs* of associated data inputs. Using the actual number of inserted event inputs, the vertical position *YPos* is calculated first. Following, the upper left corner of the place invariant to be inserted is calculated and inserted by applying the rule *insert_evarc/3* (l. 11-13). Afterwards, the transition number and their position is retrieved from the dynamic knowledge base, by proving the fact *transition/6* with a provided transition name (l. 14-15). Thereafter, the event input arc from the created event input to the transition **_Set* is inserted (l. 16). From line 17 to 20, the condition input arc between the condition input *Resource_Idle* and the transition **_Release* is inserted. As the last step, the event arcs to the transition *+_toTrue* and *+_toFalse* are inserted for every member of the list *Withs* (l. 21-23). Concluding, the rule *insert_eventinput/4* is applied to the tail of the list of event inputs to be transformed. Thereby, the number of already inserted event inputs is increased (l. 24). If the tail of the list is an empty list, the rule at line 4 is proven to be true, and the list *NextEvent* of the last transformation step is unified to an empty list. Afterwards, all calling rules are traversed back, and the list *NextEvent* holds all event inputs of the new NCE module .

Source Code 5.2 Function Block Transformation - Event Inputs

```
1  %EventInputs
2  %TNameS ... Transitionsname of the Set Transition
3  %TNrS   ... Transitionnumber
4  insert_eventinput([],_,_,[]):- !.
5  insert_eventinput([(Name, Withs)|T], Nr, XPos,
6                    [element('Event',['X'=XPos,'Y'=YPos,'Num'=Nr,'LocNum'=Nr,'Name'=Name,'
                       Comment'='_'],[])|NextEvent]):-
7          format('------------------ EventInputs --------------------~n'),
8          format('EventInput ~w~n',[Name]),
9          YPos is (Nr-1) * 150,
10         XPos2 is XPos + 50,
```

```
11          YPos2 is YPos - 50,
12          NextNr is Nr+1,
13          insert_event_var(Name, XPos2, YPos2),
14          atom_concat(Name,'_Set',TNameS),
15          transition(TNrS, TNameS, _, _, TSXPos, TSYPos),
16          insert_evarc(Name, TNrS, [(XPos,YPos), (TSXPos,TSYPos)]),
17          atom_concat(Name,'_Release',TNameR),
18          transition(TNrR, TNameR, _, _, TRXPos, TRYPos),
19          cond_in('Resource_Idle', XPos, YPosCondIn),
20          insert_condarc('Resource_Idle', TNrR, [(XPos, YPosCondIn),(TRXPos, TRYPos)]),
21          forall(member(VarName,Withs),
22                  (   format('~w ',[VarName]),
23                      insert_with(TNrS, TSXPos, TSYPos, VarName))),
24          insert_eventinput(T, NextNr, XPos, NextEvent).
```

After the transformation process has finished, the user is prompted to choose a file to store the new created *Document Object Model* of the TNCEM file. Furthermore, the module structure is represented at the left hierarchy browser. By activating the popup menu, each module can be drawn into a new frame as can be seen in Figure 2.4 in Section 2.4. This allows to check visually the transformation results, and to export the graphical representation of the NCE module into a postscript file or to copy it to the Windows owned clipboard and use it for documentation purpose at other applications.

Figure 5.2 shows the principal graphical layout of the transformation results of basic function blocks at the example of the Task-Controller of the Jack Station presented in Figure 3.6. At the top left from left to right are the NCE structures of the algorithms. Also, at the top left from top to down are the event inputs modelled. Below the NCE structures of the transformed event inputs are the data inputs modelled. The same is done for the event and data outputs at the right of the NCE module. Above the event in- and outputs, the scheduling interface is inserted, and at the middle of the module the transformed Execution Control Chart can be found. The graphical positions of the transformed EC states and EC transition are taken from the source file.

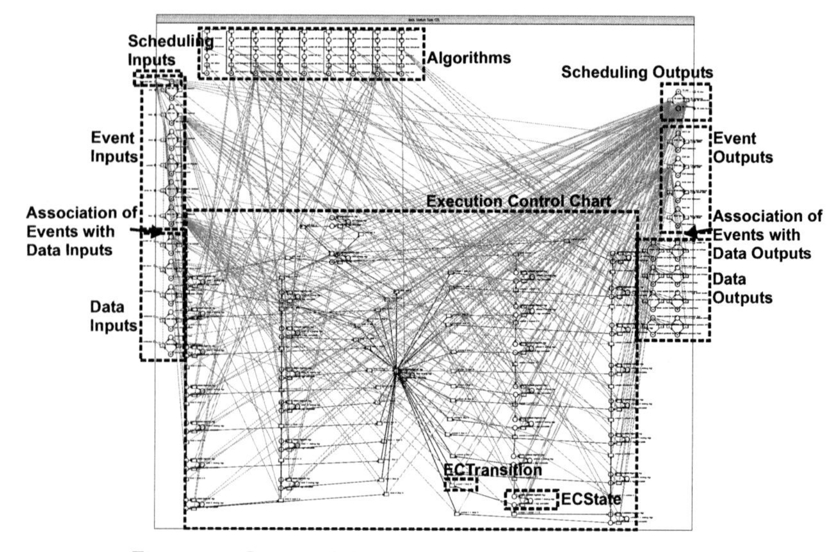

Figure 5.2 Layout of the transformation results of a basic FB

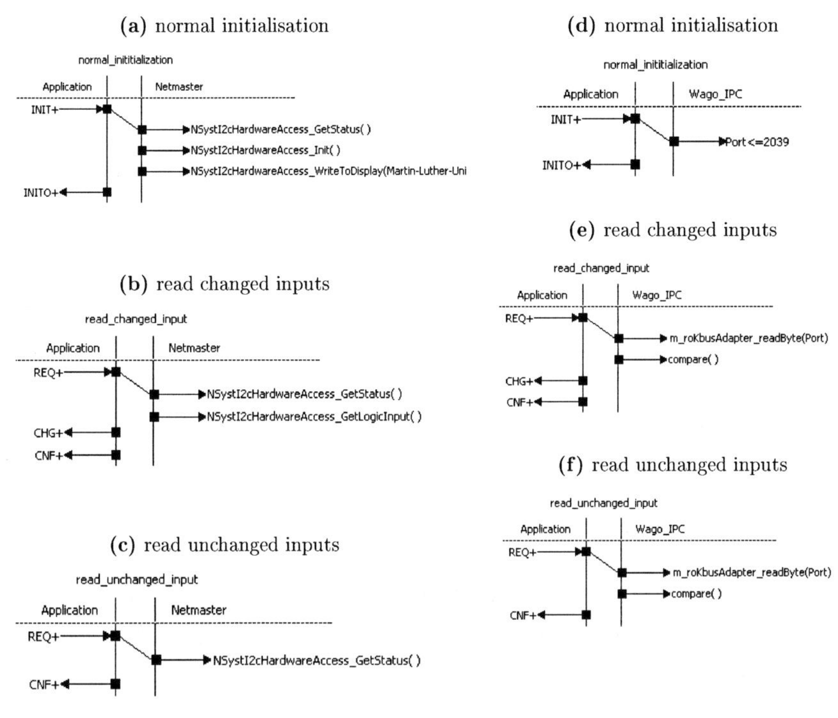

Figure 5.3 Service Sequences of the SIFB reading the inputs

5.2 Model of the Control Device

After successfully transforming the simple, basic and composite function blocks used at the control application the resource and device models are created. Therefore, the process and communication interface realised by SIFBs have to be modelled. Due to the fact that they provide a service from the underlying operation system, they do not have Execution Control Chart. Furthermore, the developers may want to hide their own intellectual property and thus the implementation is encapsulated. Also, the control engineer using the function block is in most cases not interested in the exact implementation, but only the temporal interface behaviour. Due to this reason, the standard advices to use for documentation purpose the conventions for the definition of OSI services defined in [ISO-IEC-10731]. Thus, the formal models of the SIFBs can only be derived from the provided service sequences. Figure 5.3 presents the used service sequences to document the the SIFBs to read the digital inputs of a device. From (a) to (c), the sequences for the normal initialisation process, the read of changed input values and the read of unchanged input values are presented for the *Netmaster II* used at the author's workgroup laboratory. At the right sight of Figure 5.3 the same service sequences are shown for the *Wago IPC 750-860*.

For both implementations, the sequence of the normal initialisation starts with the input event *INIT* in combination with a positive (true) event qualifier *QI* from the application. Afterwards, the implementation for the *Netmaster II* checks the hardware status to get

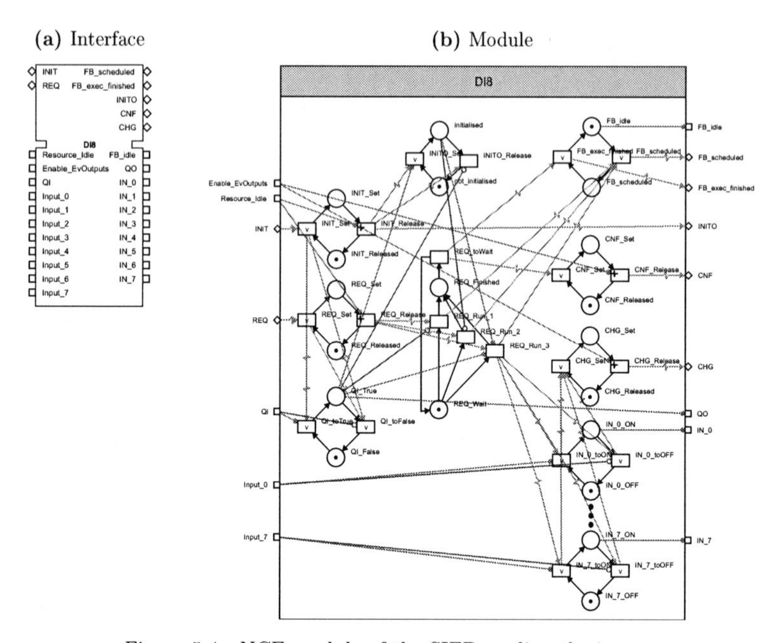

Figure 5.4 NCE module of the SIFB reading the inputs

an information, if it is already initialized. If not, the hardware is initialized, and the text *Martin-Luther-University* is written to the display. After that, the *INITO* event with a positive (true) event qualifier *QO* is published to the application. Despite this, the SIFB implementation for the *Wago IPC 750-860* checks only if the provided *Port* is inside the range of 0..2039, because the process image size is limited to 2040 bytes [WAG06]. After that, the output event *INITO* and the event qualifier *QO* equal to true is published to the application. The initialisation of the internal *KBus* and the regular update of the process image is done by another thread created during the instantiation of the SIFB. Each instance has a member variable named *m_ roKbusAdapter*, holding the unique instance of the thread to access the process image.

To read the actual input value, the application sends the event *REQ* in combination with the positive event qualifier *QI*. At the implementation of the *Netmaster*, the status flag is checked again and if it signals a change, the logic input value is retrieved, and the output events *CHG* and *CNF* are sent back to the application with the event qualifier *QO* equal to true. In contrast to this, the SIFB at the *Wago IPC 750-860* retrieves the actual value of input state from the process image and compares it with the last value. If the value has changed, the new one is provided at the data outputs, and the events *CHG* and *CNF* are published to the application with a positive event qualifier *QO*. In the fact the physical input values have not changed, only the output event *CNF* is published to the application with a positive event qualifier *QO* as shown at the lower service sequence.

Using these service sequences, an NCE module describing the causal behaviour can be developed as shown in Figure 5.4. Since, the hardware of the device is not part of the model, the *normal initialisation* service sequence can be transformed to the event input *INIT* and the event output *INITO* as well as the data input *QI* and data output *QO*. Another effect

of not incorporating the hardware into the model is, *QO* will always have the same value as *QI*. Thus, both variables are modelled by the same place invariant. Therefore, it is possible to issue the event *INITO*, during the release of the modelled *INIT* event and set the place invariant storing the information about the hardware initialisation to *Initialized*. The service sequences to read unchanged (Figure 5.3b and 5.3e) or changed (Figure 5.3c and 5.3f) input values are started from the application by sending a *REQ* event in combination with an positive event qualifier *QI*. Depending on the value of *QI* and the hardware initialisation status, the transitions *REQ_Run_1, REQ_Run_2* or *REQ_Run_3* are fired next and set the output event *CNF*. Only the transition *REQ_Run_3* forces the modelled place invariants *IN_ **, to update their value to the actual sensor state of the plant model. If at least one transition fires, the event output *CHG* is set, and accordingly to the service sequences, the modelled output events *CNF* and *CHG* are released in one of the next steps, due to the instantaneous transitions.
Concluding the model of the process interface, mainly the same analysis of the provided service sequences for the SIFB setting the output values have to be done next. Mainly the same result as for previously described SIFBs reading the input values are gained, but there is no output event *CHG*, the condition inputs are named *OUT_ **, and the condition outputs are named *Output**.
Also the standard defines in Appendix A the event handling blocks *E_Restart* and *E_Delay* by the use of service sequences, because they have to be implemented for every runtime environment separately. The function block *E_Restart* issues the events *Cold, Warm* and *Stop* to the function block network mapped to this resource, if the resource or rather the device is started, restarted or stopped. The function block *E_Delay* issues the event *EO*, at the parametrized time *DT* after the input event *START*. The timer is cancelled at the occurrence of the input event *STOP* and all multiple input events *START* are neglected.
Using the previously transformed function blocks and the NCE modules derived from the service sequences as well as a fitting resource model, a formal model of the device can be created by applying transformation rule 5 (Page 66). At the last step, all unconnected inputs and outputs of the used submodules have to be connected to the interface of the device module. For the process interface, this should be the condition inputs *Input** of the NCE modules representing the SIFBs reading the input values and the condition outputs *Output** of the NCE modules modelling the SIFBs accessing the physical outputs. The result of such an operation is shown in Figure 5.5. As done at the previous examples, all event arcs representing an event arc between function blocks are red coloured, and all condition arcs representing a data arc are drawn in blue. The green modules are the derived modules for the function blocks *E_Restart, E_Delay* and *E_Merge*, and the orange ones represent the transformed SIFBs. Thus, the green and orange NCE modules and their interconnection describe the general structure of a device. This modules and their interconnections can be used as the starting point to model any device to be verifified. Therefore, the NCE modules have to be extended by a module from Figure 4.16 fitting the scheduling function of the used runtime, as well as by the transformed function blocks mapped to this resource from the control application (blue coloured).

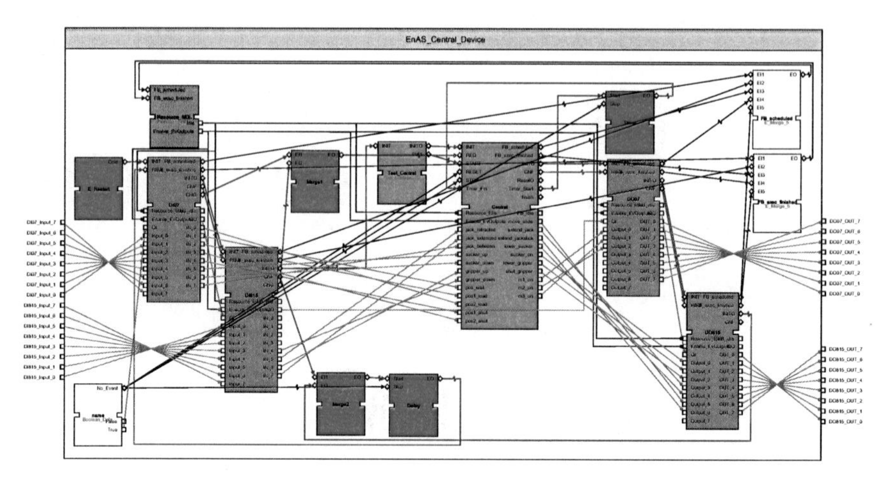

Figure 5.5 NCE module modelling the control device B of the EnAS-Demonstrator incorporating the Central Controller

5.3 Plant Model

Using the library of basic plant modules presented at section 2.5 it is feasible to derive from the structural description of the used mechanical components in Figure 5.6 the hierarchical structure of the composed formal plant model. Thus, the $_D$TNCE module describing the uncontrolled behaviour of the Gripper Station will incorporate two basic modules of the type cylinder, two of the type valve and two of the type sensor as well as an additionally local workpiece model. The Jack and the Seldge Station will use the same basic module types with a changed amount. The $_D$TNCE module describing the uncontrolled behaviour of a conveyor uses instances of the actuator module to describe the relay and the drive as well as instances of the sensor module to model the conveyor and the mounted position sensors.

Afterwards, the composite $_D$TNCE modules describing the Gripper, Jack and Seldge Station as well as the conveyors are combined to the $_D$TNCE module of each plant part. Both module instances are extended by the workpiece model of the pallet to develop the hole plant model. In the following, the signal interconnection between the used modules are discussed.

5.3.1 Gripper Station

Figure 5.7a shows the $_D$TNCE module of the Gripper Station. Depending on the *true* and *false* state of the control output, the connected pneumatic valve switches to *ON* or *OFF*. This enables the flow of compressed air into or out of the corresponding cylinder, which extends or retracts. Doing so, the gripper moves up and down or is closed and opened again. According to the duration between the opening of the valve and the start of the cylinder movement, the modelling is done with a condition interconnection. The up and down movement will turn the positioning sensors immediately to *ON* or *OFF*, if the end positions

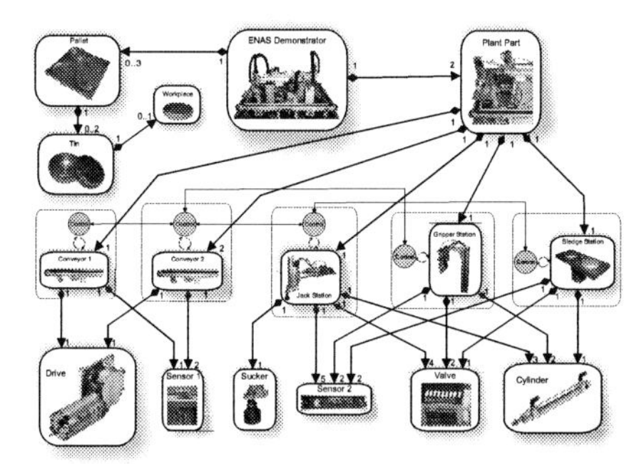

Figure 5.6 Structural description of the EnAS-Demonstrator

are reached. If a closed gripper starts moving up (↝ gripper_UpDown_not_extended) or finishes its down moving (↝ gripper_UpDown_extended), a tin will immediately be taken from or deposited to the pallet. For modelling this synchronous behaviour, event chains are used. Despite this, a tin is only closed correctly after the gripper is closed for a certain period of time, which results in modelling the connection between the closing cylinder and the tin attribute by a condition chain (—• gripper_close_not_extended). The EnAS demonstrator handles three different pallets transporting zero, one or two tins with different loading states. This workpiece behaviour of each pallet is modelled inside another $_D$TNCE module , to make the *gripper* module reusable.
For the construction purpose, the *close* cylinder is always faster than the *UpDown* cylinder, even though the valves are opened at the same time. Thus, the model of the *UpDown* cylinder is extended by discrete times, higher than the one of the close cylinder. Due to the fact of using module instances with the same temporal and causal behaviour, it is possible to parametrize the different times for extending and retracting a cylinder at the parameter file of a TNCEM or TNCES file (Section 2.4).
The local workpiece behaviour is modelled in the same way as a sensor. Thus, only one tin can be gripped at the same time, and it will be lost if the close cylinder starts opening (↝ gripper_close_not_extended). It will be gripped, if a pallet with a tin is positioned right, and the cylinder *UpDown* starts retracting in combination with an extended *close* cylinder. This decision is modelled within the global workpiece model of the pallets and will be true if an event is passed to the input *wp_toON*.

5.3.2 Jack Station

The Jack Station consists of 5 sensors, 4 valves, 3 cylinders and one sucker. Figure 5.7c shows the corresponding module. According to the explanations of the Gripper Station, the connections between the valves and cylinders are modelled by condition chains and between the cylinders and sensors by event chains. The sucker is modelled in the same way as a valve, but it will immediately suck in a workpiece and lose it, if it is switched on (↝ jack_sucker_toON) or off (↝ jack_sucker_toOFF). Accordingly, the interconnection to the local workpiece behaviour is modelled with event chains.

5.3.3 Slide Station

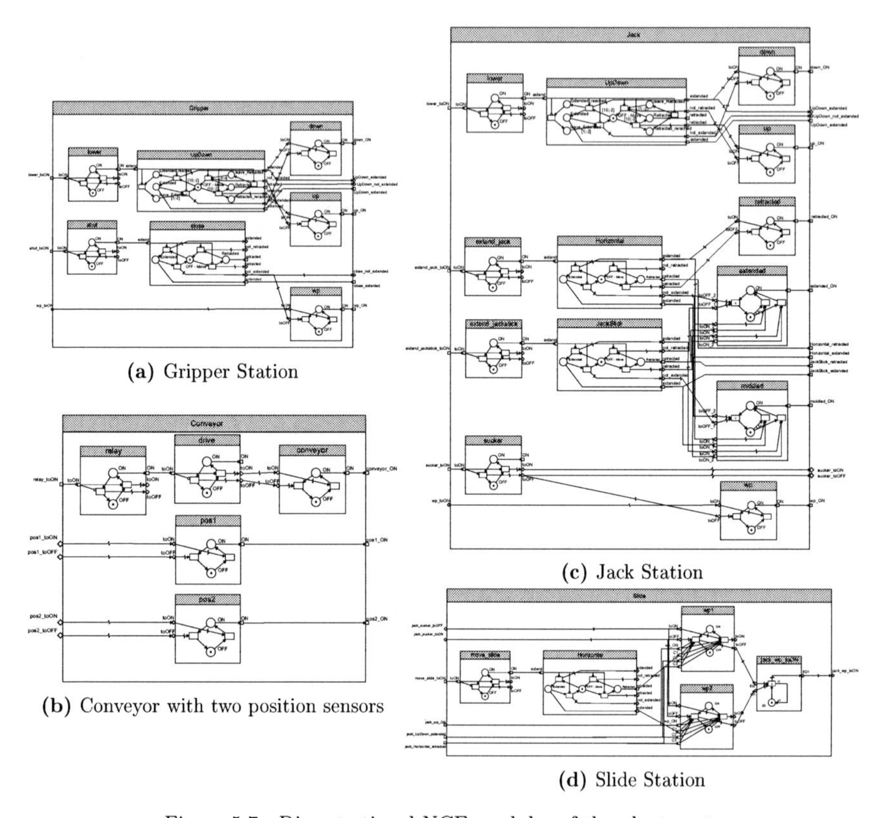

(a) Gripper Station

(b) Conveyor with two position sensors

(c) Jack Station

(d) Slide Station

Figure 5.7 Discrete timed NCE modules of the plant parts

The Slide Station modelled in Figure 5.7d consists of a valve to move a horizontal cylinder forward and backward and provides workpiece slots to the Jack Station. Into both slots, only one workpiece can be stored at the same time. Thus, modelling is done with a place invariant to toggle between the states *ON* and *OFF*, if the events *jack_sucker_toOFF* or *jack_sucker_toON* occur. Additionally, several constraints have to be fulfilled, namely a retracted *Horizontal* cylinder (—• *jack_Horizontal_retracted*) and an extended *UpDown* cylinder (—• *jack_UpDown_extended*) of the Jack Station on the one hand and an extended or retracted *Horizontal* cylinder of the Slide Station on the other hand. Moreover, a workpiece can only be sucked in or stored correctly if there is not already one. To synchronize the local workpiece models of the Jack and Slide Station, event chains are used.

5.3.4 Conveyor

Each conveyor consists of a drive with a corresponding relay to switch it on and off and the conveyor itself. Due to the mechanical connection between the drive and the conveyor, the conveyor will instantly start moving if the drive starts to rotate. According to this, the modelling is done with event chains (⇝ engine_toON, ⇝ engine_toOFF). Otherwise, it takes some time until the drive starts rotating due to the friction of itself and the conveyor. Thus, the modelling is done with a condition chain. The corresponding module is shown in Figure 5.7b.

5.3.5 Workpiece Behaviour (Pallet)

Each workpiece has several properties that are modelled globally or locally. Since pallets are moving through the whole plant, the position modelling is done in a global module interacting with all plant parts. Each station model is extended by a local workpiece model, to identify if something is sucked in, gripped or stored. The synchronization between local and global models is done by event chains, because if something is stored to the global workpiece model, it will no longer be available within the local model. Each workpiece attribute is modelled with a separate $_D$TNCE module, describing all possible states. For the pallets, the actual position as well as the loading state of the two tins on it is considered, as shown in Figure 5.8.

As presented at the left bottom in Figure 5.8, the model of the pallet position incorporates the places describing certain exact positions as the start and the end of a conveyor and the positions of the sensor as well as places describing positions between them. Only the post-arcs of the places describing a position between have time intervals assigned, to model time duration elapsing between the pallet moves from the one to the other exact position. The only condition for the pallet moving is a switched on conveyor, modelled by a condition chain.

Even if the model of the tin loading incorporates less possible states than the model of the position, it is more complicated as can be seen at the right bottom of Figure 5.8. The problem is the modelling of all the constraints, which have to be fulfilled to get from one to another state. For example to get from the state *open empty* to *workpiece sucked in*

- the pallet has to be in front of the Jack Station,
- the Jack Station has to be extended,
- the Jack Stick has to be retracted and
- the sucker has to have a workpiece sucked in,

and only if the sucker finishes now its moving down, the state is changed. The mentioned constraints are modelled as condition (—•) or inhibitor chains (—◦) to the transition, and the finally required action of finishing the moving down is modelled by an event chain (⇝) to the transition connecting the places. In the following, all state changes between the different loading changes should be examined in more detail, and the modelling within an $_D$TNCE module is derived.

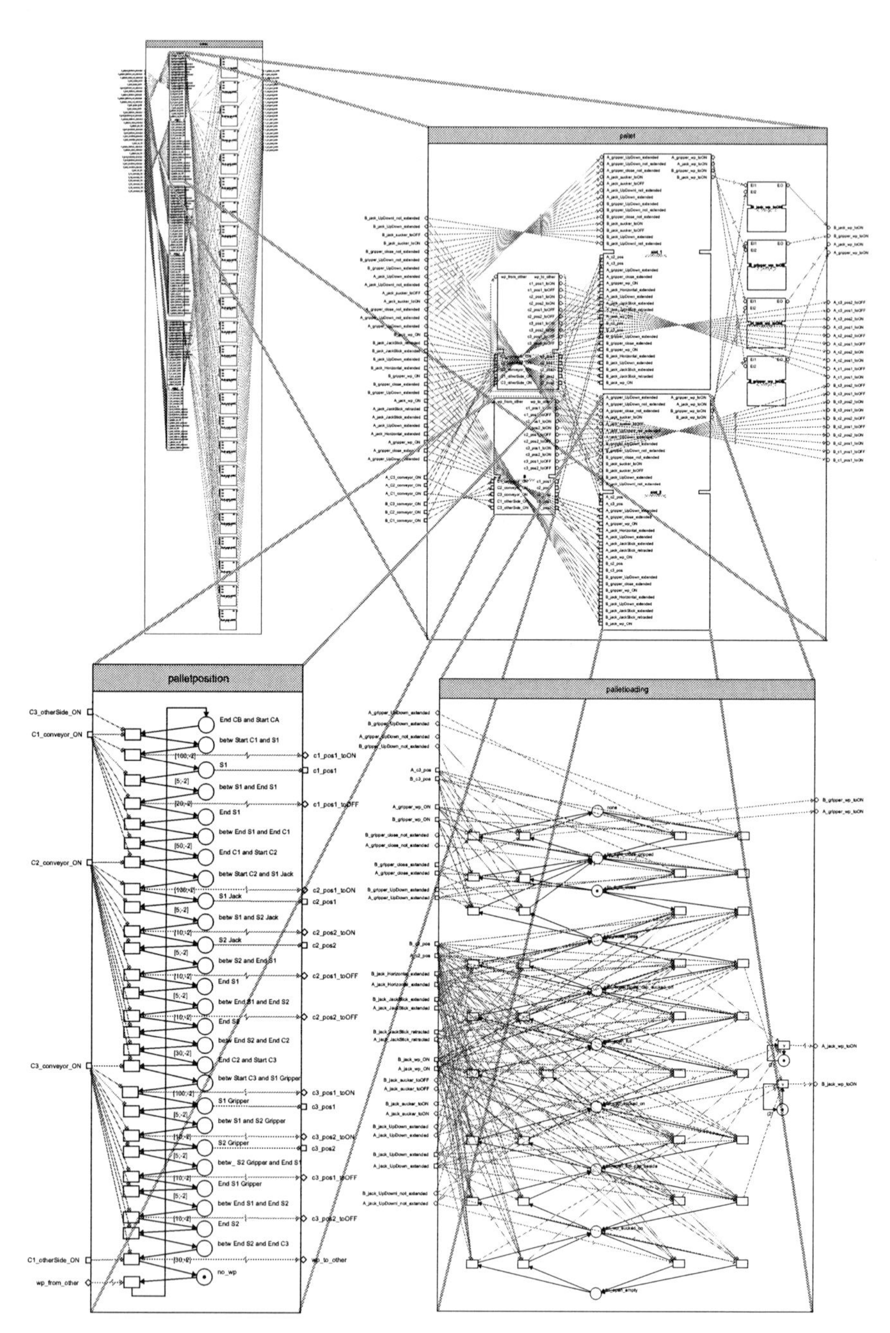

Figure 5.8 NCE module of the Pallet

Open empty: One of the possible starting states is an *opened empty* tin, which means that the tin is completely unloaded and can be loaded by the Jack Station.
It is left if the sucker of the Jack Station finishes its down moving with a sucked in workpiece.

Figure 5.9 open empty

Modelling: open empty → □ → workpiece sucked in

↝ jack_UpDown_extented
—• [jack_wp_ON, c2_pos, jack_Horizontal_extended, jack_JackStick_retracteded]

The current state is reached from *workpiece sucked in*, if the sucker starts moving up and is still switched on and holding the workpiece.

Modelling: workpiece sucked in → □ → open empty

↝ jack_UpDown_not_extented,
—• [jack_wp_ON, c2_pos, jack_Horizontal_extended, jack_JackStick_retracteded]

Workpiece sucked in: The sucker of the Jack Station is still at the lower position and has sucked in a workpiece.
If it starts moving up the previous state is reached, and if the sucker is switched off the state *open full lid beside* is reached. The constraints of a positioned pallet, an extended horizontal cylinder and a retracted Jack Stick are still the same.

Figure 5.10 workpiece sucked in

Modelling: workpiece sucked in → □ → open full lid beside

↝ jack_sucker_toOFF,
—• [jack_wp_ON, c2_pos, jack_Horizontal_extended, jack_JackStick_retracteded, jack_UpDown_extended]

This state is reached from *open full lid beside*, if the sucker is switched on.

Modelling: open full lid beside → □ → workpiece sucked in

↝ jack_sucker_toON,
—• [c2_pos, jack_UpDown_extended, jack_Horizontal_extended, jack_JackStick_retracteded, jack_UpDown_extended],
—∘ [jack_wp_ON]

Open full lid beside: After the sucker is switched off, the workpiece lies at the bottom of the tin and the Jack Station is moving to the lid beside.
If the sucker of the Jack Station is above the lid and switched on again, the state is left to *lid sucked in*.

Figure 5.11 open full lid beside

Modelling: open full lid beside → □ → lid sucked in

↝ jack_sucker_toON,
—• [c2_pos, jack_Horizontal_extended, jack_JackStick_extended, jack_UpDown_extended],
—∘ [jack_wp_ON]

The state *open full lid beside* is reached from *lid sucked in*, if the sucker is switched off.

Modelling: lid sucked in → □ → open full lid beside
⇝ jack_sucker_toOFF,
→ [c2_pos, jack_Horizontal_extended, jack_JackStick_extended, jack_UpDown_extended, jack_wp_ON]

Lid sucked in: The lid is sucked in or released, if the sucker is switched on or off. Referring to the model, the previous state is reached or left.
The successor state *open full* is reached, if the sucker of the Jack Station starts moving up with a sucked in workpiece.

Figure 5.12 lid sucked in

Modelling: lid sucked in → □ → open full
⇝ jack_UpDown_not_extended,
→ [c2_pos, jack_Horizontal_extended, jack_JackStick_extended, jack_wp_ON]

To get from the state *open full* to the current state, the sucker of the Jack Station has to reach the lower position.

Modelling: open full → □ → lid sucked in
⇝ jack_UpDown_extended,
→ [c2_pos, jack_Horizontal_extended, jack_JackStick_extended, jack_wp_ON]

Open full: This state is active as long as the lid is transported by the Jack Station.
It is reached from the previous one, if the sucker leaves the lower position and both horizontal cylinders of the Jack Station are extended. To switch to the previous state again, both horizontal cylinders have to be extended, and the sucker has to reach the lower position.

Figure 5.13 open full

If the Jack Stick of the Jack Station is retracted and the sucker reaches its down position, the state *open full* is left to the successor state *loose close lid sucked in.*

Modelling: open full → □ → loose close lid sucked in
⇝ jack_UpDown_extended,
→ [c2_pos, jack_Horizontal_extended, jack_JackStick_retracted, jack_wp_ON]

If previously the state *loose close lid sucked in* is active and the sucker starts moving up with a sucked in lid, the actual state of the loading slot model will be activated.

Modelling: loose close lid sucked in → □ → open full
⇝ jack_UpDown_not_extended,
→ [c2_pos, jack_Horizontal_extended, jack_JackStick_retracted, jack_wp_ON]

Loose close lid sucked in: The lid is still sucked in by the sucker and the *UpDown* cylinder is at the extended position. If the sucker starts moving up, the previous state *open full* is reached, and if the sucker reaches the lower position, the actual state is reached from the previous one.

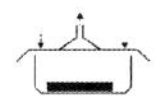

Figure 5.14 loose close lid sucked in

The sucker can now be switched off to drop the lid safely onto the tin. Thus the local workpiece model of the Jack Station is released to *no_wp* and the state of the global Workpiece Behaviour switches to the state *loose close*.

Modelling: loose close lid sucked in → loose close

⇝ jack_sucker_toOFF,
—• [c2_pos, jack_Horizontal_extended, jack_JackStick_retracted, jack_UpDown_extended, jack_wp_ON]

To get from the state *loose close* to the current state, the sucker has to be switched on.

Modelling: loose close → loose close lid sucked in

⇝ jack_sucker_toON,
—• [c2_pos, jack_Horizontal_extended, jack_JackStick_retracted, jack_UpDown_extended],
—∘ [jack_wp_ON]

Loose close: If the sucker turns off or on at its lower position, the previous state is left to the actual state *loose full* or vice versa.
The state *loose close* is active until the Gripper Station grips the tin by extending the close cylinder at the lower position.

Figure 5.15 loose close

Further on, the pallet has to be positioned at the first or second shut position of the Gripper Station, and thus the state *tight close gripped* is reached.

Modelling: Loose close → tight close gripped

—• [gripper_close_extended, c3_pos, gripper_UpDown_extended],
—∘ [gripper_wp_ON]

This state is reached from *tight close*, if the horizontal cylinder of the Jack Station is extended and the Jack Stick is retracted, and if the sucker of the Jack Station reaches the lower position.

Modelling: tight close → Loose close

⇝ jack_UpDown_extended,
—• [c2_pos, jack_Horizontal_extended, jack_JackStick_retracted]

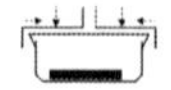

Figure 5.16 tight close gripped

Tight close gripped: As long as the gripper is closed at the lower position, the state *tight close gripped* is active.
If the gripper starts moving up, the tin is removed from the loading slot of the pallet, and the state *none* is reached.

Modelling: (tight close gripped) ⟶ [ ] ⟶ (none)

–↝ gripper_UpDown_not_extended,
—• [c3_pos, gripper_close_extended],
—∘ [gripper_wp_ON]

But if the *close* cylinder retracts, while the gripper is still at the lower position, the tin remains at the loading slot as a *tight closed* one.

Modelling: (tight close gripped) ⟶ [ ] ⟶ (tight close)

–↝ gripper_close_not_extended,
—• [c3_pos, gripper_UpDown_extended],
—∘ [gripper_wp_ON]

To reach this state from the successor *none*, the gripper has to reach its lower position with a gripped tin.

Modelling: (none) ⟶ [ ] ⟶ (tight close gripped)

–↝ gripper_UpDown_extended,
—• [c3_pos, gripper_wp_ON]

Figure 5.17 tight close

Tight close: The state *tight close* is another possible initial state of the loading slot model, and it is reached, if the gripper releases the tin by retracting the *close* cylinder at the lower position. If the *UpDown* cylinder is not extended, and consequently the gripper is not at the lower position, the tin may drop into the loading slot of the pallet but otherwise maybe not. Thus, the state *tight close* can only be reached through the state *tight close gripped*, and not directly from the state *none*.
The tin could only be opened by the Jack Station again as described at the state *loose close*.

None: Another possible initial state of the loading slot module is *none*, which means that there is no tin available, but the Gripper Station can store one.

Using the provided workpiece information, an NCE module is derived describing all possible loading states of a tin. To create an ${}_D$TNCE module describing the loading state and the position of a pallet, the tin loading module and the module describing the pallet position have to be used twice and interconnected as shown in Figure 5.8. Afterwards, the functions *extend interface* and *merge interface* provided by the TNCES-Workbench are used to extend first the interface of position modules A and B and second to merge the interface of all used submodules to create a composite module describing the behaviour of a pallet as shown at the top right of Figure 5.8. Thereby, the output events of the modules *slot_1* and *slot_2* with the module type *palletloading* are merged by several event merging modules, which incorporate a place with a capacity of 2 and a transition with the event mode $\boxed{\vee}$ at the pre- and post-set of the mentioned place. Afterwards, the semi-automatically created module describing a pallet is inserted 3 times into the global workpiece module. Again, the function *merge interface* provided by the TNCES-Workbench is used to merge the event outputs of the submodules describing a pallet, as can be seen at the top left of Figure 5.8. Thereby, the input interface of the module *pallets* is created as well and connected by event and condition interconnections to the submodules of the type *pallet*. If the number of used pallets should be increased or decreased at the real plant, the number of inserted submodules has to be changed, and the function *merge interface* of the TNCES-Workbench has to be used to create a module *pallets* with an updated number of pallets.

5.4 Summary

Each closed-loop model consists of one or more control device modules and a plant module forming an $_D$TNCE *system*. The module of the control device consists of the transformed simple, basic and composite function block gained by the automatic transformation process implemented at the TNCES-Workbench. Furthermore, several basic NCE and $_D$TNCE modules are developed at this chapter to describe the behaviour of the process and communication interface of the control device, which are implemented using SIFB and documented by service sequences. Next, the model of the plant, consisting of the mechanical components as well as the workpiece behaviour is created. Thereby, the hierarchical structure of the plant modules is derived from the composition of the components. In combination with the presented library of common plant parts in Section 2.5 and the feature of the TNCES-Workbench to *extend* or *merge* all unconnected event and condition in- or outputs of all submodules to the interface of the next higher module, it is ensured that the adaptation of the formal plant model to a changed plant configuration could be realised in a minimum of time.
To finish the closed-loop modelling of the distributed control system, the $_D$TNCE module/s of the control device/s have to be connected by condition arcs to plant module derived at the last section of this chapter. Using always the same general structure for the control device module with two SIFBs reading and writing the in- and outputs, the interconnection between the plant and the control device module has to be done once because the interface of the control device is still the same. Thus, only the module type has to be changed to verify another closed-loop system as done in the following chapter.

Chapter 6

Verfication Results

Using the closed-loop models gained with the approach described at the previous chapter, several verification results concerning the safety and process specifications as well as the influence of several scheduling functions of the resource owned function block network are presented in this chapter. The classical proof of deadlocks is neglected, because there will be only lifelocks, due to the used structure of the device module with scanning the inputs regularly. Furthermore, a production process transforms certain input products into final products, thus the initial state or any other should not be reached again, if a distributed control system of a manufacturing plant should be verified. For example, the process specification for the Central Controller of the plant part B of the EnAS demonstrator described at the beginning of Chapter 3 specifies, that a pallet has to arrive with a *thight closed* tin at position 1 and an *opened empty* tin at position 2, and it has to leave with an *opened empty* tin at position 1 and a *thight closed* tin at position 2.
Depending on the used function blocks at the device to be verified, the number of places and transition reaches 700 to 1000 easily and can climb up to several thousands. For example, the transformed function block of Master-Controller of the Conveyor has 522 places and 537 transitions, and the transformed function block of Master-Controller of the Jack Station has 444 places and 443 transitions. The most places and transitions are used to model the data input *Actions* with an integer-valued array of the size 20. Each array element is modelled by 8 place invariants consisting of 2 places and transitions as well as an additional place representing the exact value. Thus, the NCE structure of the whole array consists of $(2*8+1)*20 = 340$ places and 320 transitions. Because of the special incrementing NCE structure with 8 transitions described in Section 4.2.4, the modelling of the Algorithms, the Execution Control Chart as well as the remaining part of the interface adds roughly 100 to 200 places and transitions.

6.1 Central Controller

The transformed device module used to verify the Central Controller (Figure 3.5) of the EnAS demonstator is shown in Figure 5.5 and it is connected as described in the previous chapter to the $_D$TNCE module of the plant by condition interconnections. Due to the fact of using module instances over and over again, the initial marking and the temporal behaviour of all cylinders would be the same. Also, all pallets would have the same loading state and the same starting position. Hence, an additional parameter file has to be stored for each $_D$TNCE system or $_D$TNCE module, if it should be analysed. The parameter file used for the Central Controller is shown in Source Code 6.1. This configures the event delay between each input scan (l.2) and the start of the device (l.3) as well as the times to retract or extend a cylinder of the plant part A (l.6-11) and B (l.13-18). The starting position of pallet 2 is set to the beginning of conveyor 1 of the second plant part (l.20-21), and the loading state of the second tin is set from *tight_ close* to *open_ empty* (l.22-23).

Source Code 6.1 Used parameter file for the Central Controller

```
1  %configure device
2  external_arc_time('Device.Delay.Run', 'Device.Delay.t18', 2, -2).
3  external_arc_time('Device.E_Restart.', 'Device.E_Restart.', 70, -2).
4  external_arc_time('Device.Timer.Run', 'Device.Timer.t18', 10, -2).
5  %configure plant part A
6  external_arc_time('PLANT.A.jack.JackStick.OFF - Move', 'PLANT.A.jack.JackStick.', 13, -2).
7  external_arc_time('PLANT.A.jack.Horizontal.OFF - Move', 'PLANT.A.jack.Horizontal.', 20,
     -2).
8  external_arc_time('PLANT.A.jack.UpDown.OFF - Move', 'PLANT.A.jack.UpDown.', 10, -2).
9  external_arc_time('PLANT.A.gripper.close.OFF - Move', 'PLANT.A.gripper.close.', 3, -2).
10 external_arc_time('PLANT.A.gripper.UpDown.OFF - Move', 'PLANT.A.gripper.UpDown.', 12, -2).
11 external_arc_time('PLANT.A.sledge.horizontal.OFF - Move', 'PLANT.A.sledge.horizontal.',
     11, -2).
12 %configure plant part B
13 external_arc_time('PLANT.B.jack.JackStick.OFF - Move', 'PLANT.B.jack.JackStick.', 13, -2).
14 external_arc_time('PLANT.B.jack.Horizontal.OFF - Move', 'PLANT.B.jack.Horizontal.', 20,
     -2).
15 external_arc_time('PLANT.B.jack.UpDown.OFF - Move', 'PLANT.B.jack.UpDown.', 10, -2).
16 external_arc_time('PLANT.B.gripper.close.OFF - Move', 'PLANT.B.gripper.close.', 3, -2).
17 external_arc_time('PLANT.B.gripper.UpDown.OFF - Move', 'PLANT.B.gripper.UpDown.', 12, -2).
18 external_arc_time('PLANT.B.sledge.horizontal.OFF - Move', 'PLANT.B.sledge.horizontal.',
     11, -2).
19 %set pallet position and loading
20 external_marking('PLANT.pallets.pallet_2.B.no_wp', 0, 0).
21 external_marking('PLANT.pallets.pallet_2.B.betw Start C1 and S1', 1, 0).
22 external_marking('PLANT.pallets.pallet_2.slot_2.tin_tight_close', 0, 0).
23 external_marking('PLANT.pallets.pallet_2.slot_2.tin_open_empty', 1, 0).
```

After loading the closed-loop system into the TNCES-Workbench, the dynamic graph can be calculated using the described worker pool model in Section 2.4. At this case the graph has 10970 states and mainly a linear structure as can be seen in Figure 6.1. The calculation with only one worker at a DELL Latitude notebook with a 2GHz dual core Intel processor differs between 51 and 54 seconds, depending on the steps calculated into the depth, before the backtracking starts and the last state is put as a new source state into the message queue. Using two workers, a speed up of 28% is realised, and the calculation time differs between 36 and 39 seconds. Actually, the layout with the *dot* algorithm being part of the GraphViz package is the most time consuming calculation part with 3 minutes and 40 seconds. These times vary for each $_D$TNCE system with a similar number of reachable states, due to the different number of connecting arcs between the states. But, the mentioned times should emphasise to be able to handle the created models and to get a solution within an appropriated time. Nonetheless, the TNCES-Workbench as well as the layout algorithm of the GraphViz package can be run at an Ubuntu server with 4 CPUs or at a Solaris server with 8 CPUs via SSH and X11 tunneling or via the Thin-Clients at the computer pools of the Martin-Luther-University Halle-Wittenberg.

Figure 6.1 presents the dynamic graph for the distributed control system using the Central Controller approach. The origin with state 0 is at the bottom left and the end with state 10970 is at the top right. At the beginning, the cylinders of the plant move into their starting positions, and then the control device starts the initialisation. This means to move the pallet at conveyor 1 to the *wait_ position*.

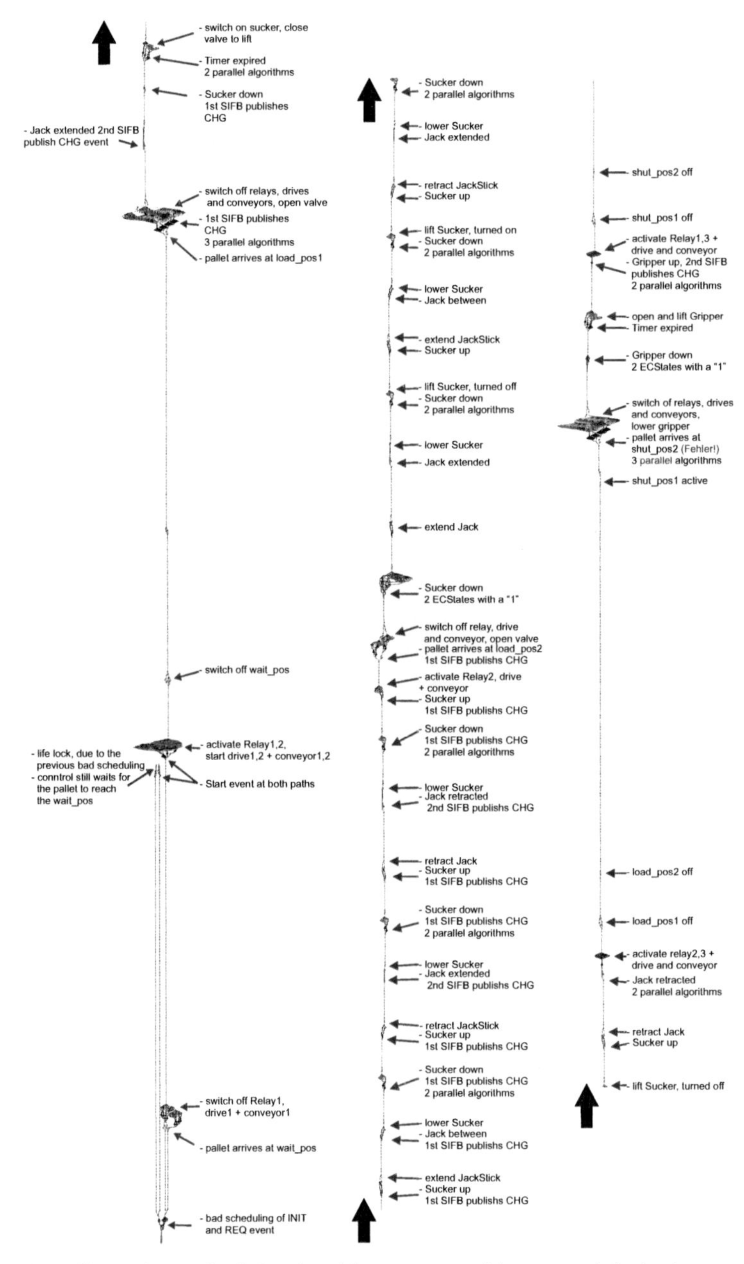

Figure 6.1 Dynamic graph of the closed-loop system with a control device incorporating the Central Controller and the EnAS-Demonstrator

As can be seen next, there exist several trajectories into a lifelock at the bottom left, which symbolise at the real system a state, where the control waits for certain sensor inputs which never get active. The trajectories leading to this lifelock have now to be analysed by examining the dynamic graph. Therefore, the TNCES-Workbench provides the possibility to visualise certain trajectories between a start and end state. Providing additionally a set of transitions (e.g. the releasing transitions *_*Release* of the modelled event inputs) a depth-first-search for all trajectories between two states can be done. Figure 6.2c presents such a trajectory from state 0 to 127, which is a pre state of the described lifelock. At the presented Gantt-Chart one can see, the blue coloured *INITO* event of the central control function block *CTL* sets the input event *INIT* of the function block *DO07*. After that, the central control function block is scheduled again and publishes the *CNF* event and sets therefore the input event *REQ* at the *DO07* function block. Now, the *DO07* function block is scheduled with the *REQ* event and publishes the *CNF* event. This sets the input event *REQ* of the function block *DO815*. But now, the *INIT* event of the function block *DO07* is scheduled. This preferred scheduling of the *REQ* event would result in a hardware exception (Netmaster II), due to a not initialised hardware before accessing it, or the function block may not be in the right state to accept the input event *REQ* (Wago IPC). This means, the physical output and the relay of the drive will not be activated, and the pallet does not start moving and never reaches the wait position. For this reason, the function block of the Central Controller hangs at the EC state *RESET*, and the $_D$TNCE system is in a lifelock scanning regularly the modelled inputs.

As can be seen at the Figure 6.2a and 6.2b, the described scenario will never be observed by using the function block execution runtimes FBRT or FORTE, because the first uses *common function calls* and the other one a *First-In-First-Out* queue to propagate the events

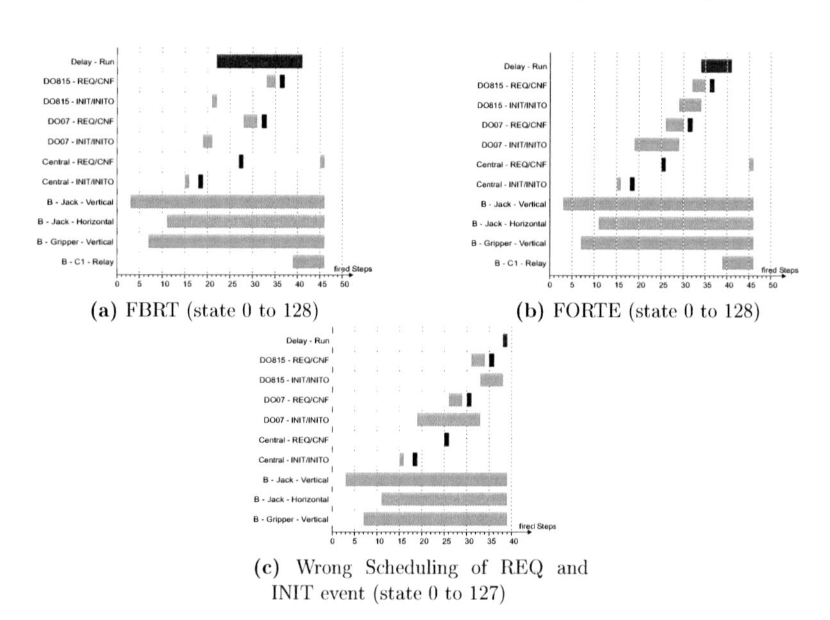

(a) FBRT (state 0 to 128)

(b) FORTE (state 0 to 128)

(c) Wrong Scheduling of REQ and INIT event (state 0 to 127)

Figure 6.2 Different Schedules during the initialisation of the control application

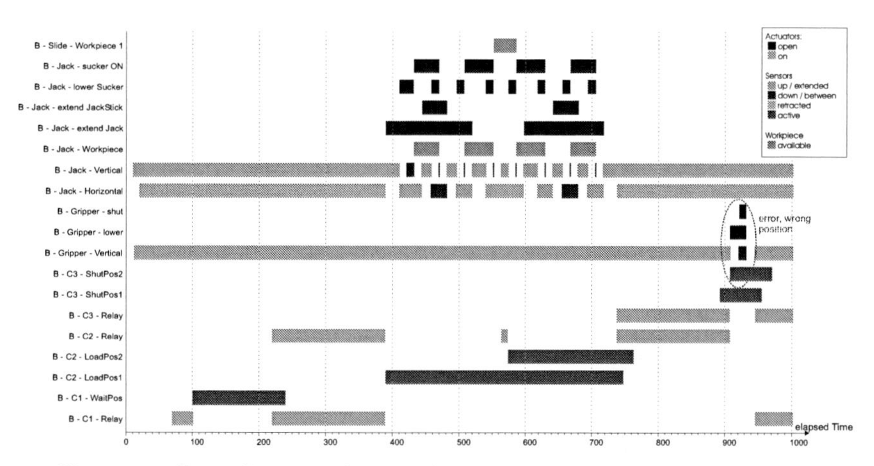

Figure 6.3 Gantt-Chart visualisation of a trajectory from state 0 to state 10970

between function blocks. This ensures that the service interface function blocks *DO07* and *DO815* are scheduled first with the *INIT* event and later with the *REQ* event, thus the following event chains are received:

FBRT: ... CTL.INITO $\rightarrow$ DO07.INIT, DO07.INITO $\rightarrow$ DO815.INIT, DO815.INITO $\rightarrow$ Merge2.EI2, Merge2.EO $\rightarrow$ Delay.Start, CTL.CNF $\rightarrow$ DO07.REQ, DO07.CNF $\rightarrow$ DO815.REQ ...

FORTE: ... CTL.INITO $\rightarrow$ DO07.INIT, CTL.CNF $\rightarrow$ DO07.REQ, DO07.INITO $\rightarrow$ DO815.INIT, DO07.CNF $\rightarrow$ DO815.REQ, DO815.INITO $\rightarrow$ Merge2.EI2 ...

Thus, the left branch can be neglected at the ongoing verification process focused at the runtimes FBRT and FORTE, but it has to be kept in mind if the function block application is mapped to a device with another runtime, not using *common function calls* or a *First-In-First-Out* queue.

For the ongoing verification of the right branch, certain steps are coloured differently. For example, all steps occurring at the plant model are *green* coloured, and all steps fired at the input scan are *brown* coloured. Depending on the number of parallel executed algorithms and which service interface function block detects the sensor change, several structures appear over and over again at the graphical representation of the dynamic graph, which eases the verification as presented in Figure 6.1. Also it could be seen, that all trajectories lead to the same lifelock at the top right. One of them is presented at the Gantt-Chart in Figure 6.3, which is easier to analyse by a control engineer than the whole dynamic graph. Furthermore, the failure of closing the wrong tin can be detected faster, because the 1st and 2nd position sensor are active, which means the second tin of the pallet is position in front of the Gripper Station instead of the first one.

6.2 Master-Task-Controller

Transforming each Master and Task-Controller used at the control application of the EnAS-Demonstrator is the first, and modelling the closed-loop system is the second step to get the verification results of this control approach. Depending on the controlled machine and the desired production process, the formal model as well as the dynamic graph gets huge and maybe not manageable even at the UNIX servers of the institute. Furthermore, several structural parts of the dynamic graph will occur over and over again, because different pallets are moved to the first or second position of the conveyors and different tins will be loaded, unloaded and gripped. Thus, the global state will be different due to the different control state and different workpieces, but due to performing the same production step only with another workpiece the structure of occurring states will be the same. Also, the parametrized production process of the EnAS-Demonstrator includes 4 rounds for the 3 pallets and loading and unloading them at the left or right Jack Station as well as gripping a tin from one pallet and deposing it to another, and if always the same control actions are triggered as described in Section 3.3, the verification result should be the same the 1st, the 2nd and .... time, if the right order of control actions is chosen. Hence, this chapter should demonstrate a way to perform a layered verification as mentioned in [MHH07], but the therein presented ideas are only proven at a simplified Task-Controller model and plant model and have to be extended to be applicable in a practical manner. The main disadvantage is the verification of the *task-plant* level with an open input interface, which means

1. the input events can occur at any state of the closed loop of the plant and the Task-Controller, which causes at least a doubling of the dynamic graph for each input event.
2. the input conditions *Resource_Idle* and *Enable_EvOutputs* from the scheduling model have to be analysed for each possible value, which would lead to an explosion of the dynamic graph.

Thus, several assumption about the environment surrounding the control loop of the Task-Controller and the controlled mechanical component have to be made and included into the verification process. If afterwards the environment consisting of the Master-Controller, the resource model and the global workpiece behaviour does not fulfil this assumptions, the model of the coordination level will have a deadlock.
The new proposed method to do a layered verification is as follows:

1. Design the formal models of each mechanical component with local and if necessary the global workpiece behaviour.
2. Transform the control function blocks.
3. Formulate the safety and process specification as well as the necessary assumptions about the environment behaviour (upper/coordination level, resource model, global or other local workpiece behaviour) for each control loop consisting of
    - the Task-Controller and the plant at **1st iteration** and
    - the Master-Controller and the reduced models at **2nd iteration**.

4. Design several $_D$TNCE modules fulfilling the assumptions and connect them to the control loops.
5. Verify each control loop separately.
6. Perform a model reduction for each control loop according to the open in- and output interface to the environment.
7. *go to the upper/coordination level and start with 2nd*

Step 3 - Formulate the safety and process specification as well as the necessary assumptions about the environment behaviour for each control loop

Steps 1 and 2 of the proposed method are already described in Section 5.3 and Section 5.1. The environment elements *resource model* and *global workpiece model* also exist already, and they are described in Section 4.3 and Section 5.3 and can be used instead of other assumptions. Thus, only some assumptions about the coordination level have to be made. As described in Section 3.3, the coordination level is responsible for coordinating all Task-Controllers by triggering certain actions by input events and waiting until their completion (output events) before triggering the next action of the same Task-Controller. Thereby, the following assumptions can be made for the control loop of Task-Controller of the conveyor and the conveyor itself:

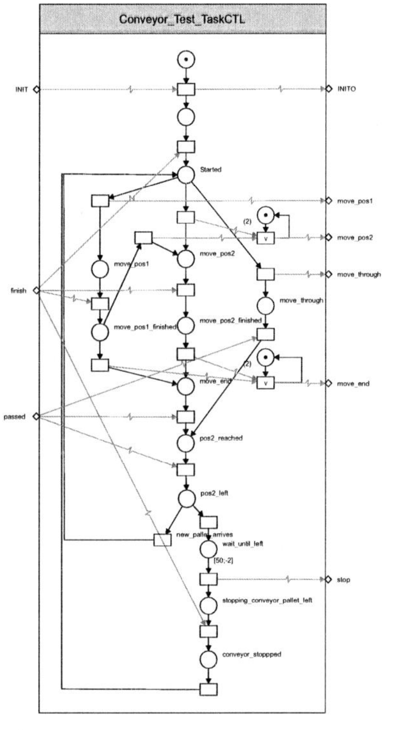

Figure 6.4 NCE module resulting from the assumption of the coordination level

- After sending the input event *Stop* the conveyor is stopped.
- After sending the input event *move_pos1* or *move_pos2* wait until the occurrence of the output event *finish*.
- After sending the input event *move_end* or *move_through* wait until the second occurrence of the output event *passed*.

Nonetheless, it is senseless in the meaning of a production progress to send the input event *move_pos1*, if the prior triggered action was already *move_pos1*, because the pallet is already there and the corresponding sensor *position_1* is active. Accordingly, the same can be applied to *move_pos2*. To achieve a production progress, it is useful to send one of the following event sequences:

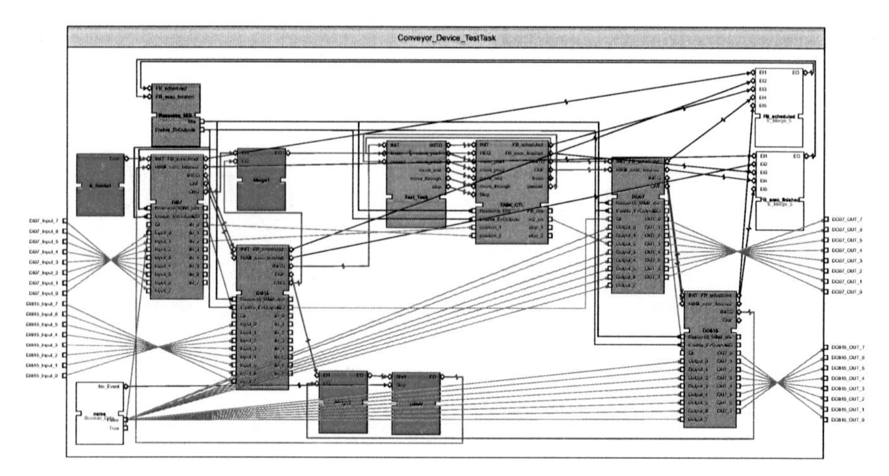

Figure 6.5 Control device model used to verify the Task-Controller of the conveyor

1. move_pos1 - move_end – (brown coloured at the dynamic graph),
2. move_pos1 - move_pos2 - move_end – (blue coloured at the dynamic graph),
3. move_pos2 - move_end – (red coloured at the dynamic graph),
4. move_through – (green coloured at the dynamic graph),
5. Stop.

Step 4 - Design several discrete timed NCE modules fulfilling the assumptions and connect them to the control loops

Figure 6.4 shows an NCE module realising all of these event sequence and taking the prior assumptions into account. Connecting this module as well as the model of the scheduling function of the resource to the control loop consisting of the Task-Controller and the plant is done in Figure 6.5. Using mainly the same module instances and module interconnections as well as the same interface of the modelled device as done for the Central-Controller approach in Figure 5.5 eases the work enormously.

Step 5 - Verify each control loop separately

The blue coloured places and transitions in Figure 6.4 represent the second event sequence, and the resulting trajectory at the dynamic graph of the closed loop (Figure 6.6) is blue coloured too. State 0 can be found at the beginning of the short trajectory above the 1st scale-up, and due to the regular input scan each trajectory ends in a lifelock as previously described for the Central-Controller approach. Each coloured trajectory represents one of the four described sequences of events and starts at state 0 and ends at state 5266 (bottom left). Thereby, the blue coloured trajectory overlays all other, because it is the most complex one and all other trajectories are split from or merged to it.
The 1st scale-up shows from top to bottom the blue coloured trajectory representing the 2nd event sequence, and at the middle the action *move_pos1* is triggered. Since, the green

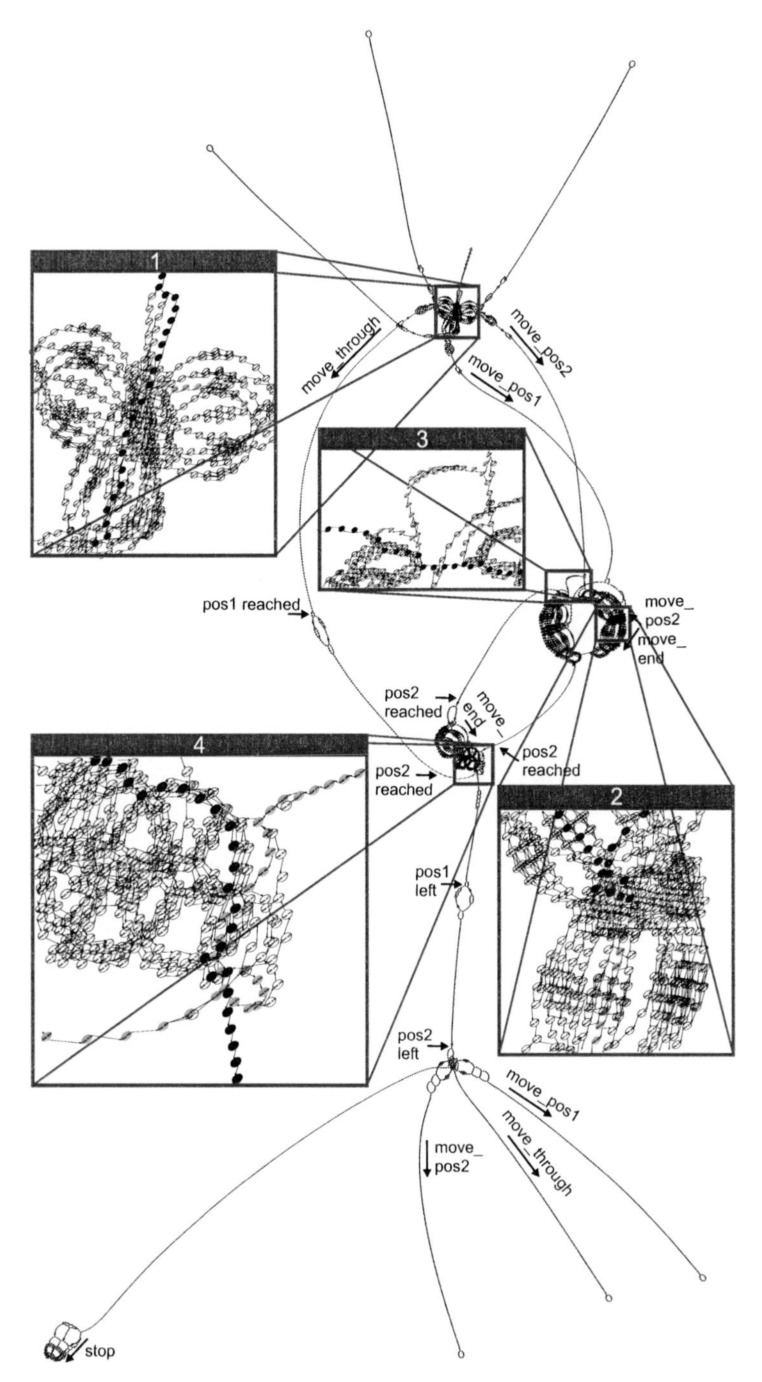

Figure 6.6 Dynamic graph of the closed-loop system

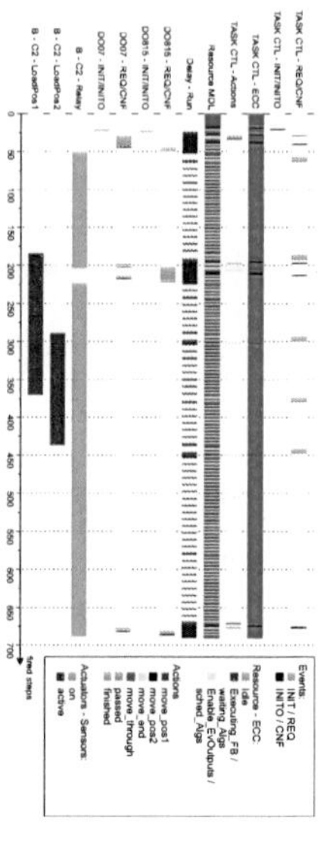

(a) Executing actions *move_pos1* and *move_end* (Trajectory 0-1782-5266)

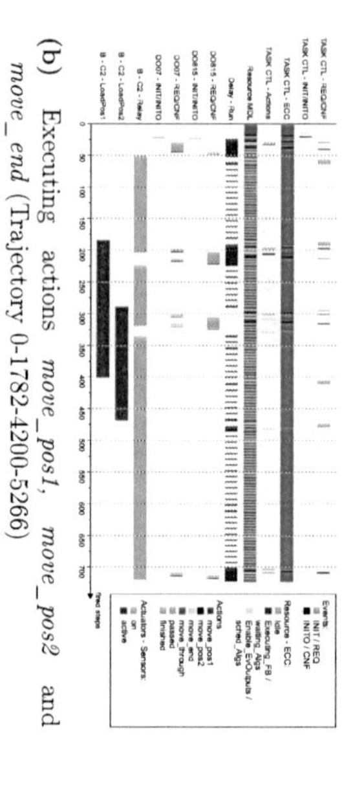

(b) Executing actions *move_pos1*, *move_pos2* and *move_end* (Trajectory 0-1782-4200-5266)

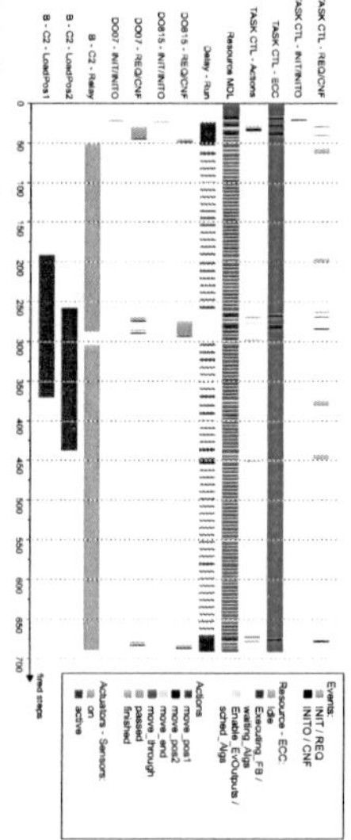

(c) Executing actions *move_pos2* and *move_end* (Trajectory 0-4200-5266)

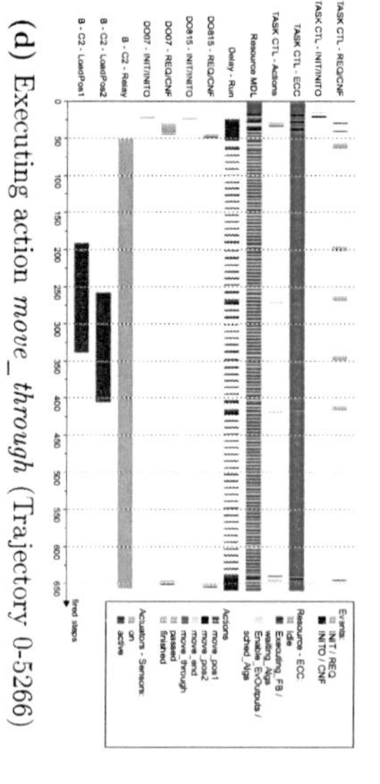

(d) Executing action *move_through* (Trajectory 0-5266)

Figure 6.7 Gantt-Chart of the different trajectories

coloured trajectory represents the 4th event sequence, the action *move_ through* is triggered and the blue and green trajectories are split at the middle of the 1st scale-up too. At the same point, the red coloured trajectory is split as well, because the action *move_pos2* is triggered.
At the 2nd scale-up, the blue trajectory gets in at the top right and leaves at the top left. At the middle of this scale-up, the brown coloured trajectory representing the 1st event sequence is split from the blue one, because the action *move_ end* is triggered instead of the action *move_pos2*. In the following, the red trajectory is merged with the blue one at the 3rd scale-up because both are moving the pallet to the 2nd position with the action *move_pos2*.
The 2nd position is reached by all trajectories before the 4th scale-up and at the brown, red and blue trajectory the action *move_ end* is executed to move the pallet to the end of the conveyor. If the 2nd position is reached by the workpiece, the actions *move_ through* and *move_ end* are merged at the Task-Controller, and thus all trajectories will be merged, if the modelled workpiece behaviour reaches the second position as shown at the 4th scale-up. After that, the 1st position is left and afterwards the 2nd one. Thereafter, new actions as *move_pos1, move_pos2* or *move_ through* can be started to move the following pallet to the desired positions. Thereby, the former pallet will leave the conveyor. If no pallet is following and no other action is started, the conveyor will be stopped by the input event *Stop*, after a certain time, and the pallet has left the conveyor.
The corresponding Gantt-Charts to the described trajectories are shown from Figure 6.7a to 6.7d in the same order as the event sequences were defined at Page 96. In contrast to the Gantt-Chart visualisation of the Central-Controller, the x-axis represents now the number of *fired steps* instead of the *elapsed time*. This means even if a bar is thicker than another, the elapsed time may be the same or even lower. For example, during each blue coloured bar at the *Delay - Run* line, the time 2 elapses. This is the time between each new input scan of the control device done by two service interface function blocks. According to this, the resource model switches between the states *Idle* (green), *Executing_ FB* (red) and *Enable_ EvOutputs* (yellow). At the line *TASK CTL - Actions*, the sequence of the received input events triggering the execution of an action can be checked, and at line *TASK CTL - ECC* the behaviour of the *ECC operation state machine* defined at the standard. According to transformation rule 3.1 at Page 47, this state machine exists for every EC state and not only one for the ECC, but as can be seen, the behaviour is the same with the states *Idle* (green), *scheduling Algorithms* (yellow) and *waiting Algorithms* (red). The two top lines show the occurrences of the event pairs *INIT/INITO* and *REQ/CNF*. Each time a sensor change is detected by the service interface function blocks scanning the inputs, a *REQ* event occurs at the interface of the function block *TASK_ CTL*, but the *ECC operation state machine* switches only into the state *scheduling Algorithms* if an EC transition clears.
At the three lower lines the actual sensor an actuator states of the conveyor are shown. This enables the control engineer to decide easily, if the action iss performed correctly. If the Gantt-Chart is extended by the modelled pallet position, the control engineer can decide as well, if the pallet is exactly positioned above the sensor and not too far away.

Step 6 - Perform a model reduction for each control loop according to the open in- and output interface to the environment

At this step an equivalent $_D$TNCE module has to be created for the control loop consisting of the Task-Controller and the plant. In [DJL02] it is defined that two modules

are equivalent if they have the same in- and output structure $\Psi(S_T, \Phi)$ and the same in- and output behaviour. The set Φ of in- and outputs is defined by the in- and outputs the environment is connected to, and the behaviour of the in- and outputs is described by the the calculated dynamic graph. This graph has to be reduced with the condition to include the information of the workpiece property and the position as well as the sensor and actuator state, because production processes serve for workpiece transformation and flow as well as the actuator and sensor state. That will be reflected in the coordination layer specifications and therefore has to be part of the model. Furthermore, it will ensure that the control engineer is able to easily understand the presented counter example or to check if the verification results are true.
As starting point of the reduction the reduction rules of [Kar09] can be used, but they have to be extended to include the elapsed time as well. The set of interesting transitions is the set of incomplete controlled transitions if the previously connected environment is removed from the model.
This reduced ${}_{D}$TNCE module can be used as described in [MHH07] for the verification of the coordination layer, but before the steps 3 to 6 have to be repeated for all control loops including a Task-Controller and a mechanical component.

6.3 Workpiece Controller

The method of a layered verification used to for the Master-Task-Controller at the previous section can be applied for the Workpiece Controller as well. Once again, the 1st and 2nd step is already described in Section 5.3 and Section 5.1, and next several assumptions about the environment of the elementary control loops have to be formulated. As done at

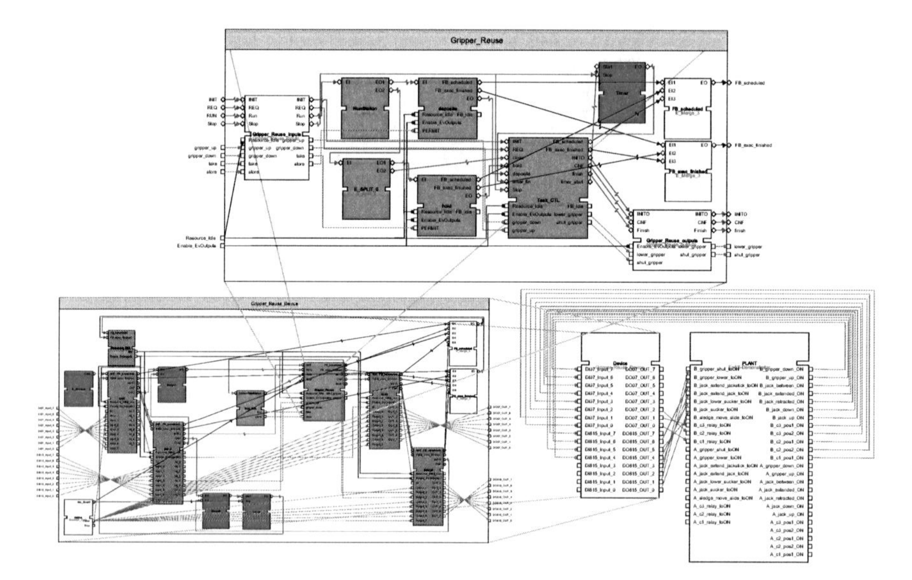

Figure 6.8 Closed-loop system to verify the control of the Gripper Station

the verification of the Master-Task-Controller, the models of the resource and the global workpiece behaviour can be used instead of any other assumptions. Thus, only assumption about the interaction with the higher level have to be made, which are as follows:

- Send *Run(store=false, take=true)* to take a tin from the pallet and wait until the event output *finish*, before sending *Run(store=true, take=false)* to put a tin to a pallet and wait until the event output *finish*.
- Send *Run(store=false, take=false)* to close a tin and wait until the event output *finish*.
- Do not send *Run(store=true, take=true)*.

Using these assumptions, an $_D$TNCE module can be created and interconnected as shown in Figure 6.8 to the control loop incorporating the the Gripper Station. The last assumption has to be proven at the next higher level for the modelled control loop, because it could not be modelled or it has to be changed to:

- Send *Run(store=true, take={true; false})* to put a tin to a pallet

if the improved function block network of Figure 3.11 at Page 38 is used to control the Gripper Station.

The dynamic graph presented in Figure 6.9 of the control loop includes 5893 states, and the source state 0 is at the middle. Once again, every trajectory ends in a live lock, and if the REQ event of the modelled service interface function blocks is scheduled before the INIT event, no modelled output of the device gets active, and the control gets into a lifelock during executing an action. The upper part of the graph shows the behaviour of the control loop if it gets the input event *REQ* in combination with the data inputs *store* and *take* set to *false* (2nd sequence). Thus, the gripper moves down to close a tin and moves up again without a tin. At the lower part, the behaviour of the control loop is presented for executing the 1st defined sequence. This means *a tin should be lifted and hold until it is put down again.* But as can be seen, there exist several trajectories according to the scheduling function used by the resource, which are not merged together. The ones going to the right includes the scheduling of the Forte runtime, and the others going to the left includes the FBRT. The both resulting trajectories intersect below, and as can be seen from the layout structure of the graph and from the scheduling Gantt-Charts, the one (Forte) performs the action to *close*, and the other (FBRT) to *hold* a tin, depending on the release of the input event *close* or *hold* first. Thus, only similar scheduling of function blocks to the FBRT control the plant in the appropriated way and fulfil the process specification *to lift a tin.* After the gripper is up again, the output event *finish* is published to the environment, and the environment sends back *Run(store=true, take=false)* to put the lifted tin back to a pallet (dotted line). At this point, the scheduling of the function blocks results again in two separated trajectories as described above. The trajectories going to the left at the left scheduling point include the first-in-first-out scheduling used by the *Forte* runtime, and once again the action *close* is executed. The trajectories going to the right from the left scheduling point and the trajectories going to the left from the right scheduling point are executing the action *deposite*, but the left one with no previously lifted tin and the other at the right with a lifted tin. Therefore, the trajectories are merged at the point, where the tin is put again to the pallet and the gripper moves up. All these mentioned trajectories end in the same lifelock, when the gripper is up again.

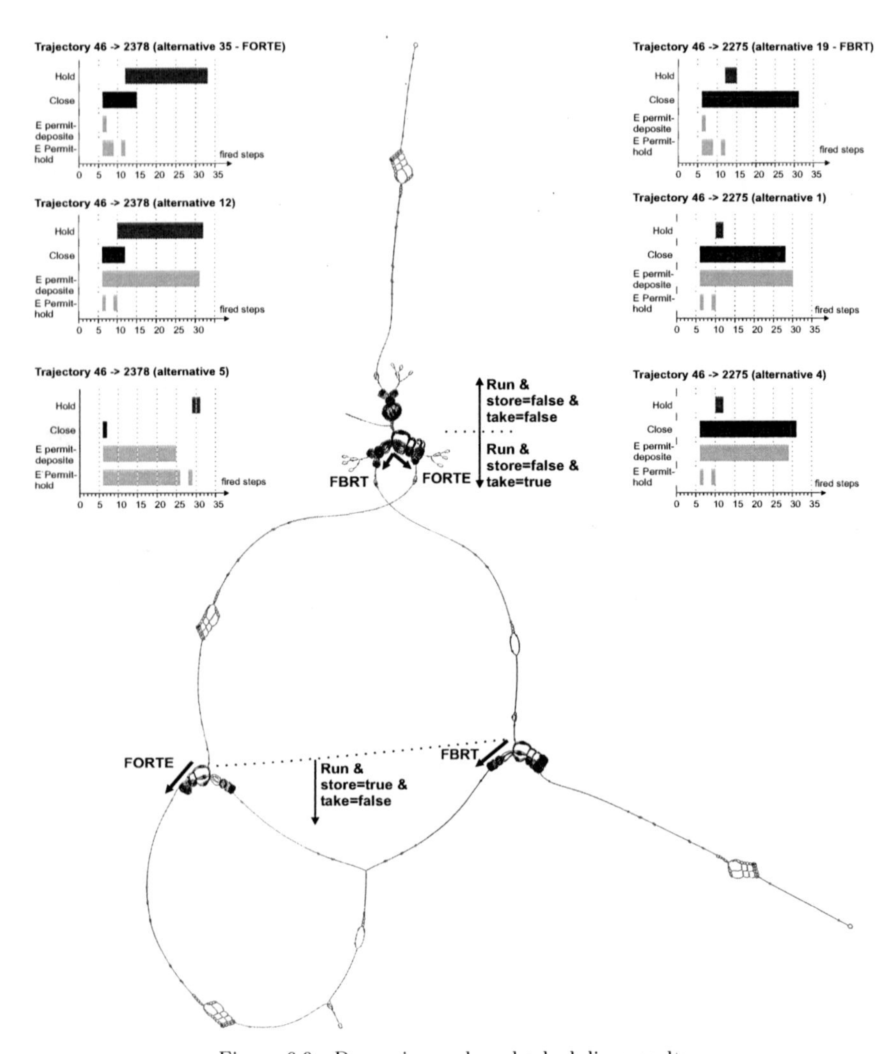

Figure 6.9 Dynamic graph and scheduling results

Nonetheless, there exists a critical scheduling possibility of the function blocks, which may cause damage to humans and the machine. At the second scheduling point it is possible to trigger the *close* action (all trajectories going to the right), if previously a tin was gripped. Thereby the gripper is opened, before it is in the lower position (states drawn red), and the tin drops down. As long as the handled workpieces are only tins with approximately 200g, nothing dangerous will happen, but at a real production plant it could lead to dangerous situations. Furthermore, this critical situation can only be reached, if there exists a non-deterministic scheduling as mentioned in [VDVF07], but to the author's best knowledge such a runtime does not exists, but nonetheless this scheduling possibility leading to a dangerous situation has to be kept in mind.
Furthermore, this critical situation could be avoided if the improved function block network of Figure 3.11 at Page 38 is used, as described in [GH10]. Due to the use of cascaded event switcher, the event qualifier *store* is dominant and therefore the action to put a tin to the pallet.
Now, the behavioural model described by the dynamic graph can be reduced and an equivalent $_D$TNCE module created, with the same open inputs and outputs to the environment and the same input output behaviour. This step works as described in the previous chapter and afterwards the higher level can be verified.

6.4 Summary

Using the formal models of the closed-loop systems consisting of the device modules gained from the control approaches described in Chapter 3, and the plant module developed from a library of often used plant components at the end of Chapter 5, reachability analysis is performed. Starting with the approach of the Central Controller each dynamic graph is analysed with the TNCES-Workbench providing features as searching for forbidden states, finding all trajectories between two or more states and taking the different order of a provided transition set into account as well as visualising a trajectory as Gantt-Chart. Due to the fact that the scheduling function of the resource is included to the model, a lifelock is found by all control approaches, if the *REQ* event is scheduled before the *INIT* event of the implemented SIFB. Because the runtimes FBRT and Forte are used during the thesis, this behaviour was never observed at the real plant, but it may happen if another runtime is used.
To cope with the state explosion, the Central Controller is verified for the plant parts A and B separately with the assumption about how the workpiece pallet has to arrive at the desired plant part. Due to the fact, it is verified for this assumption, how the pallet will leave plant part B, the assumption for the other plant part A is fulfilled and if the other plant part A is verified the same way, than the assumption for B is fulfilled as well. Thus the distributed control system is verified.
Using this approach of separating the verification for each control loop, a method is described and used for the verification of the Parametrized Master-Task-Controller. Therein, all control loops consisting of a Task-Controller and the corresponding plant parts are verified first and than a reduction is performed to the behaviour of the unconnected in- and outputs. This reduced models are used to verify the upper coordination layer of the Master-Controller.
Examining the control loop of the function block network presented in Figure 3.10 and the Gripper Station shows, that the behaviour of the Gripper Station changes from using the

runtime *FBRT* or *Forte*, due to the different scheduling functions. Using the *FBRT* the Gripper Station grips a tin and lifts it and deposites it afterwards. Using the runtime *Forte* the tin is always closed at the lower position of the Gripper Station. Thus, the function block network is improved to the one presented in Figure 3.11 and now the behaviour of the plant is the same for both runtimes.
Nonetheless, it should be mentioned that all results like dynamic graphs and visualised trajectories are gained within an appropriate time with a usual Laptop PC, but due to the fact of using the expert system SWI-Prolog, it is possible to run the TNCES-Workbench even at an Ubuntu or Solaris cluster available at the Institute of Informatics.

Chapter 7

Conclusions

The presented thesis solves several open issues concerning the formal modelling of distributed control system complying with the standard IEC 61499 and the verification of the developed systems. As presented in the last chapter, the approach of formal modelling used in this work with a separated Execution, Data Processing and Scheduling Model is practical and feasible. Furthermore, different concepts to implement distributed control systems were presented. Nonetheless, there are new open issues that need further investigation before the verification of distributed control system complying with the IEC 61499 is ready for industrial application. These open issues concern the industrial application of the IEC 61499 itself, and the automatic process to gain the verification results and an appropriate presentation of them for the control engineer working in practice.

7.1 Summary of the thesis

As introductorily mentioned, there exist several key issues like modularity, reusability, portability, flexibility and reconfigurability for modern industrial automation. Therefore, medium-sized companies can feature themselves by hardware manufacturing independence, but using still the same basic software objects for different projects, instead of developing them every time from scratch. Due to the reconfigurability, also the plant operating company benefits from it, due to increased flexibility in the production process. However, this will also increase the complexity of the implemented distributed control system as shown by the presented control approaches. These approaches are focused to the upcoming standard IEC 61499, because in the view of the author it is better to program a control application first, and distribute it to the available devices later. Each control application consists of function blocks, which exchange information about state changes by event connections and data values by data connections. But as already reported in literature, there exist critical interconnections as *feedback loops* and *event splittings* due to the use of different scheduling functions for the components of a resource. Each of the available scheduling functions has another advantage like the best performance for an application with mainly a parallel or sequential order of function blocks. Another scheduling function is easy to implement at an already existing *Programmable Logic Controller* or to be used with *Field Programmable Gate Arrays*. Thus, each of them enriches the IEC 61499, but the plant behaviour has to be examined for the different scheduling functions before a control application is ported from one runtime to another one, which was the aim of this thesis.

Therefore, the syntax and semantic definitions of the *discrete timed Net Condition/Event Systems* are revised first, and a markup language is described to exchange once developed $_D$TNCE modules between different software tools. The revision of the definition leads to a re-implementation of the used model checker in the expert system SWI-Prolog. During the ongoing work, the resulting TNCES-Workbench is extended by several modules as for example the automatic function block translator and the visual representation of trajectories as Gantt-Charts, which are easier to understand by a control engineer than a set of fired steps.
Starting with a short introduction into the software architecture of an IEC 61499 complied distributed control system, several implementation approaches are presented for a testbed available at the Chair for Automation Technology. The first one incorporates one monolithic and therefore not reusable function block controlling one plant part and is named *Central Controller*. But, grouping the EC states and EC transitions concerning different mechanical components together, a *Task-Controller* for each component of the plant is developed. These Task-Controllers are used by the following *Master-Task-Controller* approach and coordinated by the *Master-Controller*. Thus, only the coordination layer has to be updated if the production process or the plant configuration is changed. But due to the fact that not every runtime is able to perform an online update of function blocks, the idea of implementing all production scenarios is realised with the approach of *Parametrized Master-Task-Controller*. It can be switched between the different production scenarios by providing another parameter set, consisting of an array of actions to be performed. The main drawback of this control implementation approach is the development of a production scenario, which gets even worse if more and more pallets are introduced to the system. Thus, the concept of the *Workpiece Controller* is developed. Each pallet is controlled by one Workpiece Controller, and this one incorporates the planed production scenario of the pallet. It allocates all necessary Task-Controllers for the next production step and releases the ones that are not needed any longer. Despite the approaches used for manufacturing plants, also a control approach from the field of process control is presented.
To provide the basis for an automatic function block translator and the formal modelling of an IEC 61499 system configuration, an Execution, a Data Processing and a Scheduling Model are defined next. The Execution Model incorporates the formal modelling of the interface with event in- and outputs as well as Boolean and integer-valued data in- and outputs. Additionally, the transformation of arrays using the already mentioned elementary data types is defined too. Furthermore, the Execution Model incorporates the model of the EC transitions and their conditions as well as the EC states and their assigned EC actions and a general structure of the algorithms. This general structure of the algorithms is the connecting point between the Execution and the Data Processing Model. The Data Processing Model is derived from the already proven methods of informatics and includes several NCE structures to add, subtract, increment, decrement and compare integer-valued variables. Furthermore, these binary modelled variables can be set to a static value or to the value of another variable. All necessary Boolean operations like set, reset, negate and compare are included as well into the Data Processing Model. The defined Scheduling Model of function block networks models the scheduling function used by the resource and incorporates the sequential and parallel execution of function blocks as well as the synchronous one and the cyclical scan of each function block. Thus, to the best knowledge of the author, all already existing function block execution runtimes can be modelled.
Using the defined models, a closed-loop modelling is performed next by transforming first all basic function blocks and then all composite ones to connect them to the model of the resource of the control device. The resource model incorporates one of the NCE modules

presented at the scheduling model, which is connected to all function block models that a resource model consists of. Now, a model for the service interface function blocks is developed to model the process and communication interface of the control device. These service interface function blocks encapsulate a service of the underlying operation system, and each vendor hides its intellectual property and documents only the behaviour of the interface. But due to the fact that the vendor ensures this behaviour, only this is of interest for the ongoing verification process and can be used for formal modelling of the service interface function blocks. As presented for the service interface function block reading the physical inputs, the NCE modules for all other service interface function blocks can be developed. Finally, all the unconnected in- or outputs of the developed NCE modules are connected by condition interconnection to the interface of the control device module. The plant module will be connected to the interface of the created control device modules. The model of the plant is developed by using a predefined library of $_D$TNCE modules for often used plant components. Thus, the hierarchy of the derived $_D$TNCE modules corresponds to the structure of the plant. Only the model of the workpiece has to be developed for each plant from scratch.
Each closed-loop model is analysed by calculating the dynamic graph and checking it against a provided production and safety specification by using the TNCES-Workbench. Due to the fact of analysing all possible schedules, it is observed that all presented control approaches incorporate a lifelock, where the control would wait for certain sensor input, which will never occur. At the real system this is never observed, because the used runtimes FBRT and Forte do not schedule the *REQ* event of a service interface function block before the *INIT* event. All false schedules would lead to not activating the actuators, because the service interface function block is in the false state, and the *REQ* event is withdrawn. To cope with the state explosion, the Central Controller of each plant part is verified separately with the assumption, how the workpiece pallet has to arrive. Thus, it could be verified how the pallet will leave the plant part, and if this fulfils the assumption of the arriving of the pallet for the other plant part, the 2nd one can be verified. After that, it is ensured how the pallet arrives again at the 1st plant part, which fulfils the assumption as well. Therefore, the verification succeeded for both plant parts. The approach of verifying each control loop separately is extended for the Master-Task-Controller concept as well as for the Workpiece controller. During the verification of the control loop consisting of the Gripper Station and its control, different plant behaviours are detected by using the runtimes FBRT and Forte. Using the *FBRT* the Gripper Stations moves down, grips a tin and lifts it, and deposites it afterwards. Contrary to this the Gripper Station always moves down and close a tin, if the *Forte* runtime is used.

7.2 Ongoing work

Before getting the verification of distributed control system into industrial application, the development of them has to get it. Therefore, the development and implementation of distributed control system in combination with the IEC 61499 has to be part of the lectures at the university as done already by Chair for Automation Technology. Beyond that, it has to be included in the lectures and practical courses at the university of applied science as well as to the lessons and industrial placements at the professional schools, because the new control engineers working in practice are educated there. This could be achieved by training the teaching personal to the already commercial available engineering environments from the *nxtControl GmbH* and *ICS Triplex ISaGRAF Inc.*.

To increase the demonstration facilities concerning distribution, reusability and portability, the existing testbed should be equipped with different control devices from different vendors. Each of them should be equipped with a webserver realised as a service interface function block and providing the visualisation of the controlled plant components. By sending the requested website back to the requesting client, it should be updated with the current data of the control application. Because a website can be presented with any browser, created by anyone who has even no knowledge of the concepts of the IEC 61499 and linked easily to other websites at other devices, this will be a suitable solution for distributed visualisation.
To increase the acceptance of the new standard by industrial partners, it will be useful to transform step by step all IEC 61131 function blocks of the OSCAT library [1] to function blocks of the IEC 61499. Thus, the generally used function blocks would be available, and if a partner would like to switch or try to test, he will not need to start from scratch. Furthermore, the formal models of these function blocks can be provided as well.
As already described at the beginning of Section 4.2.6, there is only a limited set of the Structured Text included to develop the Data Processing Model, but in the ongoing work, step 4 of the *MatIEC* compiler developed at a collaborative project between the *University of Porto* and the *TBI SARL - Lolitech* and being part of the open source framework Beremiz, should be changed to create the presented NCE structures of the Data Processing Model. Doing this, the lexical analyses and parsing the syntax as well as a semantic check is already available for the programming languages *Instruction List, Structured Text* and *Sequential Function Chart*, and the generation of the NCE structures has to be programmed only once for all of them.
Next, the reduction of the dynamic graph should be improved, and a model of the communication channel between the devices should be introduced into the formal model. But the communication between the control devices will be the next higher level above the coordination layer consisting of Master-Controller and the reduced models of the Task-Controller and the controlled plant components.

7.3 Long-term vision

To improve the verification it may be necessary to use another model checker like SMV or NuSMV, but therefore the vertically and horizontally composed *discrete timed Net Condition/Event Systems* have to be transformed into the input language of the tools, and each counter has to be transformed back to the $_D$TNCE systems. As the starting point the work of [Wim97] and the results of a DAAD sponsored exchange by Prof. Victor Dubinin can be used.
Furthermore, if the counter examples of the model checker are reduced to a sequence of sensor and actuator changes, it may be possible to include this sequence into modelling tools like *Enterprise Dynamics* and to observe the counter example at a 3D model of the plant, which may be easier to understand than a Gantt-Chart. Also, this 3D plant model can be used in advance to the verification for a closed-loop simulation interconnected with the real hardware, and only if the simulation satisfies a verification is performed. Last but not least, the 3D model of the plant can be created from the already existing CAD model, because most of the simulation tools as for example *Enterpise Dynamics* are able to import VRML or 3ds files, which can be exported from most CAD tools as AutoCAD, SolidWorks, SolidEdge and Pro/Engineer.

[1] `http://www.oscat.de`

References

References to Journals and Conference Proceedings

[BVT+08] R.W. Brennan, P. Vrba, P. Tichy, A. Zoitl, C. Suender, T. Strasser, and V. Marik. Developments in dynamic and intelligent reconfiguration of industrial automation. *Computers in Industry*, 59(6):533 – 547, 2008.

[CA08] G. Cengic and K. Akesson. A Control Software Development Method Using IEC 61499 Function Blocks, Simulation and Formal Verification. In *Proceedings of the IFAC World Congress (IFAC)*, volume 17, pages 22–27, Seoul, Korea, 2008.

[CB06a] J. Chouinard and R.W. Brennan. Software for next generation automation and control. In *Proceedings of the Conference on Industrial Informatics (INDIN)*, pages 886–891, Singapore, 2006.

[CC06] K.-H. Cheng and S.-W. Cheng. Improved 32-bit Conditional Sum Adder for Low-Power High-Speed Applications. *Journal of Information Science and Engineering*, 22(4):975–989, 2006.

[CCB06] M. Colla, E. Carpanzano, and A. Brusafferri. Applying the IEC61499 model to the shoe manufacturing sector. In *Proceedings of the Conference on Emerging Technologies and Factory Automation (ETFA)*, pages 1301–1308, September 2006.

[CLA06] G. Cengic, O. Ljungkrantz, and K. Akesson. Formal Modeling of Function Block Applications Running in IEC 61499 Execution Runtime. In *Proceedings of the Conference on Emerging Technologies and Factory Automation (ETFA)*, pages 1269 – 1276, Prague, Czech Republic, 2006.

[DBCT07] G. Doukas, A. Brusaferri, M. Colla, and K. Thramboulidis. RTAI-based Execution Environments for Function Block Based Control Applications. In *Proceedings of the Conference on Emerging Technologies and Factory Automation (ETFA)*, pages 1489 – 1496, Patras, Greece, 2007.

[DJL02] J. Desel, G. Juhás, and R. Lorenz. Input/output equivalence of petri modules. In *6th Biennal World Conference on Integrated Design and Process Technology (IDPT)*, Pasadena, California, USA, 2002.

[DT08] G. Doukas and K. Thramboulidis. Implementation Model Alternatives for IEC 61499 Function Block Networks. In *Proceedings of the Conference on Industrial Informatics (INDIN)*, pages 295–300, Daejeon, South Corea, 2008.

[DV08] V. Dubinin and V. Vyatkin. On Definition of a Formal Model for IEC 61499 Function Blocks. *EURASIP Journal on Embedded Systems*, 2008 (Article ID 426713):10, 2008.

[DVH06] V. Dubinin, V. Vyatkin, and H.-M. Hanisch. Modelling and Verification of IEC 61499 Applications using Prolog. In *Proceedings of the Conference on Emerging Technologies and Factory Automation (ETFA)*, pages 774–781, Prague, Czech Republic, September 2006.

[FV04] L. Ferrarini and C. Veber. Implementation approaches for the execution model of IEC 61499 applications. In *Proceedings of the Conference on Industrial Informatics (INDIN)*, pages 612–617, Berlin, Germany, June 2004.

[GH10] C. Gerber and H.-M. Hanisch. Does portability of IEC 61499 mean that once programmed control software runs everywhere? In *Proceedings of the 10th IFAC Workshop on Intelligent Manufacturing Systems (IMS)*, pages 29–34, Lisbon, Portugal, July 2010.

[GHE08] C. Gerber, H.-M. Hanisch, and S. Ebbinghaus. From IEC 61131 to IEC 61499 for Distributed Systems: A Case Study. *EURASIP Journal on Embedded Systems*, 2008:8, 2008.

[GHH09] C. Gerber, M. Hirsch, and H.-M. Hanisch. Automatisierung einer energieautarken Fertigungsanlage nach IEC 61499. *Automatisierungstechnische Praxis (atp)*, 51(03):44–52, 2009.

[GIVH08] C. Gerber, I. Ivanova-Vasileva, and H.-M. Hanisch. A Data processing Model of IEC 61499 Function Blocks with Integer-Valued Data Types. In *Proceedings of the 9th IFAC Workshop on Intelligent Manufacturing Systems (IMS)*, pages 239–244, Szczecin, Poland, October 2008.

[GIVH10] C. Gerber, I. Ivanova-Vasileva, and H.-M. Hanisch. Formal modelling of IEC 61499 function blocks with integer-valued data types. *Control and Cybernetics*, 39(1):197 – 231, 2010.

[GKNV93] E. Gansner, E. Koutsofios, S. North, and K.-P. Vo. A technique for drawing directed graphs. *IEEE Transactions on Software Engineering*, 19(3):214–230, 1993.

[GPH10] C. Gerber, S. Preuße, and H.-M. Hanisch. A Complete Framework for Controller Verification in Manufacturing. In *Proceedings of the Conference on Emerging Technologies and Factory Automation (ETFA)*, pages 1–9, MF-001279, Bilbao, Spain, September 2010.

[HGVH07] M. Hirsch, C. Gerber, V. Vyatkin, and H.-M. Hanisch. Design and Implementation of Heterogeneous Distributed Controllers according to the IEC 61499 Standard - A Case Study. In *Proceedings of the Conference on Industrial Informatics (INDIN)*, pages 829–834, Vienna, Austria, July 2007.

[HKL99] H.-M. Hanisch, P. Kemper, and A. Lüder. A modular and compositional approach to modeling and controller verification of manufacturing systems. In *Proceedings of the IFAC World Congress (IFAC)*, volume J, pages 187–192, Beijing, China, July 1999. Preprints.

[HM98] M. Heiner and T. Menzel. Instruction list verification using petri net semantics. In *Proceedings of the Conference on Systems, Man and Cybernetics (SMC)*, volume 1, pages 716–721, October 1998.

[HSZ+06] O. Hummer, C. Suender, A. Zoitl, T. Strasser, M.N. Rooker, and G. Ebenhofer. Towards Zero-downtime Evolution of Distributed Control Applications via Evolution Control based on IEC 61499. In *Proceedings of the Conference on Emerging Technologies and Factory Automation (ETFA)*, pages 1285 – 1292, Prague, Czech Republic, 2006.

[HW08] N. Hagge and B. Wagner. Analyzing the liveliness of IEC 61499 function blocks. In *Proceedings of the Conference on Emerging Technologies and Factory Automation (ETFA)*, pages 377 – 382, Hamburg, Germany, 2008.

[IVGH07] I. Ivanova-Vasileva, C. Gerber, and H.-M. Hanisch. Transformation of IEC 61499 Control Systems to formals Models. In *Proceedings of the Conference Automatics and Informatics (CAI)*, pages V–5 – V–10, Sofia, Bulgaria, October 2007.

[IVGH08] I. Ivanova-Vasileva, C. Gerber, and H.-M. Hanisch. Basics of Modelling IEC 61499 Function Blocks with Integer-Valued Data Types. In *Proceedings of the 9th IFAC Workshop on Intelligent Manufacturing Systems (IMS)*, pages 233–238, Szczecin, Poland, October 2008.

[LF80] R. E. Ladner and M. J. Fischer. Parallel prefic computation. *Journal of Association for Computing Machinery*, 27 (4):821 – 838, 1980.

[MFC99] C. Maffezzoni, L. Ferrarini, and E. Carpanzano. Object-oriented models for advanced automation engineering. *Control Engineering Practice*, 7(8):957 – 968, 1999.

[MHH07] D. Missal, M. Hirsch, and H.-M. Hanisch. Hierarchical distributed controllers - design and verification. In *Proceedings of the Conference on Emerging Technologies and Factory Automation (ETFA)*, pages 657–664, Patras, Greece, September 2007.

[OWRSB05] S. Olsen, J. Wang, A. Ramirez-Serrano, and R. W. Brennan. Contingencies-based reconfiguration of distributed factory automation. *Robotics and Computer-Integrated Manufacturing*, 21(2005):379390, 2005.

[PGH10b] S. Preuße, C. Gerber, and H.-M. Hanisch. Virtual Start-Up of Plants using Formal Methods. *Journal of Modelling, Identification and Control (IJMIC)*, 2010. Manuscript accepted for publication.

[Pie05] E. Pietriga. A toolkit for addressing hci issues in visual language environments. In *Proceedings of the Symposium on Visual Languages and Human-Centric Computing (VL/HCC)*, pages 145–152, Dallas, Texas, USA, September 2005.

[PV07] C. Pang and V. Vyatkin. Towards Formal Verification of IEC61499: modelling of Data and Algorithms in NCES. In *Proceedings of the Conference on Industrial Informatics (INDIN)*, volume 2, pages 879 – 884, Vienna, Austria, 2007.

[PV08] C. Pang and V. Vyatkin. Automatic model generation of IEC 61499 function block using net condition/event systems. In *Proceedings of the Conference on Industrial Informatics (INDIN)*, pages 1133 – 1138, Daejeon, South Corea, 2008.

[SFL02] C. Schnakenbourg, J.-M. Faure, and J.-J. Lesage. Towards IEC 61499 function blocks diagrams verification. In *Proceedings of the Conference on Systems, Man and Cybernetics (SMC)*, volume 3, page 6, Hammamet, Tunisia, October 2002.

[SG04] M.-P. Stanica and H. Gueguen. Using timed automata for the verification of IEC 61499 applications. In *Proceedings of the IFAC Workshop on Discrete Event Systems (WODES)*, pages 22–24, Reims, France, 2004.

[SRE+08a] T. Strasser, M. Rooker, G. Ebenhofer, A. Zoitl, C. Suender, A. Valentini, and A. Martel. Framework for Distributed Industrial Automation and Control (4DIAC). In *Proceedings of the Conference on Industrial Informatics (INDIN)*, pages 283–288, Daejeon, South Corea, 2008.

[SRE+08b] T. Strasser, M. Rooker, G. Ebenhofer, A. Zoitl, C. Suender, A. Valentini, and A. Martel. Structuring of large scale distributed control programs with IEC 61499 subapplications and a hierarchical plant structure model. In *Proceedings of the Conference on Emerging Technologies and Factory Automation (ETFA)*, pages 934–941, Hamburg, Germany, 2008.

[Sue08] C. Suender. *Evaluation of Downtimeless System Evolution in Automation and Control Systems.* PhD thesis, Vienna University of Technology, 2008.

[SZC+06] C. Suender, A. Zoitl, J.H. Christensen, V. Vyatkin, R.W. Brennan, A. Valentini, L. Ferrarini, T. Strasser, J.-L Martinez-Lastra, and F. Auinger. Usability and Interoperability of IEC 61499 based distributed automation systems. In *Proceedings of the Conference on Industrial Informatics (INDIN)*, pages 31–37, Singapore, China, August 2006.

[SZC+07] C. Suender, A. Zoitl, J.H. Christensen, M. Colla, and T. Strasser. Execution Models for the IEC 61499 elements Composite Function Block and Subapplication. In *Proceedings of the Conference on Industrial Informatics (INDIN)*, pages 1169 – 1175, Vienna, Austria, 2007.

[TBdS07] E. Tisserant, L. Bessard, and M. de Sousa. An Open Source IEC 61131-3 Integrated Development Environment. In *Proceedings of the Conference on Industrial Informatics (INDIN)*, pages 183–187, Vienna, Austria, 2007.

[TD06a] K. Thramboulidis and G. Doukas. IEC61499 Execution Model Semantics. In *Proceedings of the Conference on Industrial Electronics, Technology & Automation (CISSE-IETA)*, 2006.

[TZ05] K. Thramboulidis and A. Zoupas. Real-time java in control and automation: a model driven development approach. In *Proceedings of the Conference on Emerging Technologies and Factory Automation (ETFA)*, volume 1, pages 8–46, Catania, Italy, September 2005.

[VCL05] V. Vyatkin, J.H. Christensen, and J.L.M. Lastra. OOONEIDA: an open, object-oriented knowledge economy for intelligent industrial automation. *Transactions on Industrial Informatics*, 1 (1):4 – 17, 2005.

[VD07] V. Vyatkin and V. Dubinin. Sequential Axiomatic Model for Execution of Basic Function Blocks in IEC61499. In *Proceedings of the Conference on industrial Informatics (INDIN)*, pages 1183 – 1188, Vienna, Autrich, 2007.

[VDVF07] V. Vyatkin, V. Dubinin, C. Veber, and L. Ferrarini. Alternatives for Execution Semantics of IEC61499. In *Proceedings of the Conference on Industrial Informatics (INDIN)*, pages 1151–1156, Vienna, Austria, 2007.

[VH99] V. Vyatkin and H.-M. Hanisch. A Modeling Approach for Verification of the IEC 1499 Function Blocks Using Net Condition / Event Systems. In *Proceedings of the Conference on Emerging Technologies and Factory Automation (ETFA)*, pages 261–269, Barcelona, Catalonia, Spain, 1999.

[VH00a] V. Vyatkin and H.-M. Hanisch. Development of adequate formalisms for verification of IEC1499 distributed applications. In *Proceddings of the Conference of Society of Instrument and Control Engineers (SICE)*, pages 73–78, Iizuka, Japan, July 2000.

[VH00c] V. Vyatkin and H.-M. Hanisch. Modelling of IEC 61499 function blocks as a clue to their verification. In *Design and optimization of intelligent machine tools*, page 18, Karpacz, Poland, 2000.

[VH01c] V. Vyatkin and H.-M. Hanisch. Bringing the model-based verification of distributed control systems into the engineering practice. In *Proceedings of IFAC Workshop on Intelligent Manufacturing Systems (IMS)*, pages 152 – 157, Poznan, Poland, 2001.

[VH01d] V. Vyatkin and H.-M. Hanisch. Formal Modeling and Verification in the Software Engineering Framework of IEC61499: A Way to Self-verifying Systems. In *Proceedings of the Conference on Emerging Technologies and Factory Automation (ETFA)*, pages 113 – 118, Antibes-Juan les Pins, France, 2001.

[VH03] V. Vyatkin and H.-M. Hanisch. Verification of distributed control systems in intelligent manufacturing. *International Journal of Manufacturing*, 14 (1):123 – 136, Februar 2003.

[VH05b] V. Vyatkin and H.-M. Hanisch. Reuse of components in formal modelling and verification of distributed control systems. In *Proceedings of the Conference on Emerging Technologies and Factory Automation (ETFA)*, pages 129–134, 2005.

[VHH06] V. Vyatkin, M. Hirsch, and H.-M. Hanisch. Systematic Design and Implementation of Distributed Controllers in Industrial Automation. In *Proceedings of the Conference on Emerging Technologies and Factory Automation (ETFA)*, pages 633–640, September 2006.

[VKP05] V. Vyatkin, S. Karras, and T. Pfeiffer. An architecture for automation system development based on IEC 61499 standard. In *Proceedings of the Conference on Industrial Informatics (INDIN)*, pages 13–18, Perth, Australia, August 2005.

[Wie09] J. Wielemaker. *Logic programming for knowledge-intensive interactive applications.* PhD thesis, University of Amsterdam, Faculty of Science, June 2009.

[Wim97] G. Wimmel. *A BDD-based Model Checker for the PEP Tool.* PhD thesis, University of Newcastle, May 1997.

[WSS03] Z. Wang, F. H. Y. Sun, and Y. Song. Colored Petri net model of IEC function block and its application. In *Proceedings of the Conference on Emerging Technologies and Factory Automation (ETFA)*, pages 648 – 651, Lisbon, Portugal, 2003.

[WW00] H. Wurmus and B. Wagner. IEC 61499 konforme Beschreibung verteilter Steuerungen mit Petri-Netzen. In *Conference Verteilte Automatisierung*, Magdeburg, 2000.

[YRVS09] L.H. Yoong, P.S. Roop, V. Vyatkin, and Z. Salcic. A Synchronous Approach for IEC 61499 Function Block Implementation. *IEEE Transactions on Computers*, 58 (12):1599–1614, 2009.

References to Books

[BDM05] B. Becker, R. Drechsler, and P. Molitor. *Technische Informatik - Eine Einführung.* Pearson Studium, 2005.

[CSF00] R. Conrad, D. Scheffner, and C. Freytag. *Conceptual Modeling – ER 2000*, volume 1920(2000) of *Lecture Notes in Computer Science (LNCS)*, chapter XML Conceptual Modeling Using UML, pages 558–571. Springer Berlin / Heidelberg, informatik edition, 2000.

[GKN06] E. Gansner, E. Koutsofios, and S. North. *Drawing graphs with dot.* dot user Manual, January 2006.

[Gut09] O. Gutzeit. *Modellbasierte Entscheidungsunterstützung bei der Fertigung bahngeführter Materialien*, volume 5 of *Hallenser Schriften zur Automatisierungstechnik.* Logos Verlag Berlin, Berlin, Germany, 2009.

[Han92a] H.-M. Hanisch. *Petri-Netze in der Verfahrenstechnik. Modellierung und Steuerung verfahrenstechnischer Systeme.* R. Oldenburg Verlag, München, 1992.

[HZ07] B. Hrúz and M. C. Zhou. *Modeling and Control of Discrete-event Dynamic Systems.* Advanced Textbooks in Control and Signal Processing. Springer, London, UK, 2007.

[Kar09] S. Karras. *Systematischer modellgestützter Entwurf von Steuerungen für Fertigungssysteme*, volume 4 of *Hallenser Schriften zur Automatisierungstechnik.* Logos Verlag Berlin, Berlin, Germany, 2009.

[Lew01] R. Lewis. *Modelling Control Systems Using IEC 61499.* Institution of Engineering and Technology, 2001.

[Rea05] M. Readman. *Servo Control Systems 2: Digital Servomechanisms.* TQ Education and Training Ltd, 2005.

[SS04] W. Schiffmann and R. Schmitz. *Technische Informatik 1.* Springer-Lehrbuch. Springer Lehrbuch, 2004.

[Thi02] J. Thieme. *Symbolische Erreichbarkeitsanalyse und automatische Implementierung strukturierter, zeitbewerteter Steuerungsmodelle*, volume 3 of *Hallenser Schriften zur Automatisierungstechnik.* Logos Verlag Berlin, Berlin, Germany, 2002.

[Vya07] V. Vyatkin. *IEC 61499 Function Blocks for embedded and distributed Control Systems Design.* O^{3}NEIDA - Instrumentation Society of America, 2007.

[WAG06] WAGO Kontakttechnik GmbH & Co. KG, D-32423 Minden. *Wago I/O System 750, Modulares I/O-System, Linux-Feldbuskoppler 750-860*, version 1.1.1 edition, 2006.

[ZD93] M. C. Zhou and F. DiCesare. *Petri Net Synthesis for Discrete Event Control of Manufacturing Systems*, volume 204 of *The Springer International Series in Engineering and Computer Science.* Kluwer Academic Publishers, Boston, MA, 1993.

[ZF04] M. C. Zhou and M. P. Fanti, editors. *Deadlock Resolution in Computer-Integrated Systems.* MARCEL DEKKER INC, 2004.

[Zho95] M. C. Zhou, editor. *Petri Nets in Flexible and Agile Automation*, volume 310 of *The Springer International Series in Engineering and Computer Science.* Kluwer Academic Publishers, 1995.

[Zoi09] A. Zoitl. *Real-Time Execution for IEC 61499.* International Society of Automation, Research Triangle Park, NC, 2009.

[ZV99] M. C. Zhou and K. Venkatech. *Modeling, Simulation, and Control of Flexible Manufacturing Systems: A Petri Net Approach*, volume 6 of *Series in Intelligent Control and Intelligent Automation.* World Scientific Publishing Company, 1999.

References to International Standards

[IEC-61131-1] Programmable controllers - Part 1: General information, 2003.

[IEC-61131-3] Programmable controllers - Part 3: Programming languages, 2003.

[IEC-61499-1] Function Blocks for Industrial Process Measurement and Control Systems - Part 1: "Architecture", 2005.

[IEC-61499-2] Function Blocks for Industrial Process Measurement and Control Systems - Part 2: "Software tool requirements", 2005.

[ISO-IEC-10731] Information technology – Open Systems Interconnection – Basic Reference Model – Conventions for the definition of OSI services, 1994.

[ISO-IEC-15909-2] Software and Systems Engineering, High-level Petri Nets - Part 2: Transfer Format, 2005.

Index

Personal details

	Personal:
Name	*Christian Gerber*
Birthday	1980-July-17th
Place of birth	Halle/Saale
Nationality	German
Family status	Partnership, one child (birth. 2008-June-05th)

	School and Tertiary education:
1991 to 1999	secondary school "Thomas Münzer" at Halle/Saale
1999	final secondary-school examinations at German, mathematics, physics & geography
2000 to 2006	academic studies of Engineering-Informatics at the Martin-Luther-Universität Halle - Wittenberg, with the focus on process engineering and automation technology
Jan. 2006	diploma thesis: *"Erarbeiten eines verteilten Steuerungskonzeptes"* (*Diplomingenieur (Dipl.-Ing.)*)

Working experience:

2006 till now

scientific staff at the
Chair for Automation Technology,
at the Institute of Computer Science
of the Martin-Luther-University Halle-Wittenberg

- collaboration and writing the final report about the research work done by the Chair for Automation Technology at the collaborative project *"Energieautarke Aktor- und Sensorsysteme (EnAS)"* with the general management of the Festo AG & Co. KG at the BmWi supported programme *next generation media*
- developing of the working draft, financial planning as well as writing the application for 6th working package of the collaborative project "On-the-Fly-Migration und Sofort-Inbetriebnahme von Automatisierungssystemen (OMSIS)"
- developing of the working draft, financial planning as well as relisation of the industrial project "Ausarbeitung und Implementierung von Modellierungstechniken zur prediktiven Simulation hybrider Produktionssysteme in der Kunststoffindustrie" in cooperation with the firm LAE Engineering GmbH

Awards:

2010

O^{3}NEIDA Award in memory of Bernado-Ferroni at the IFAC Workshop on Intelligent Manufacturing Systems (IMS) in Lisbon with the presented paper: *Does portability of IEC 61499 mean that once programmed control software runs everywhere?*

Abstract

There exist certain key issues to modern industrial automation as modularity, reusability, portability, flexibility, extendibility and reconfigurability to create optimally coordinated automation solution for manufacturing plants. This will feature the operating companies to react fast and flexible to changed customer demands. An appropriate way to realize these issues is an object-oriented control implementation, which has been quite common since the late 90s.

This work is focused on the upcoming standard IEC 61499, which defines an object-oriented and event-driven software model, which can be realised by any hardware the engineer prefers. Furthermore, the control implementation is application-oriented and all parts are mapped to available control devices later on. Thus, it is possible to replace one device by another by easily remapping the application. But, does this possibility need certain care during the development of the control application and how could a control engineer be supported to prove the correctness of the plant behaviour in any case?

To answer this question the formal model *discrete timed Net Condition/Event Systems* is used in this work to model in a modular manner the control system and the plant. Therefore, an *Execution model* for the function blocks as well as a *Data processing model* for executed algorithm are defined. These are extended by a *Scheduling model* to examine the system behaviour, if different but standard compliant runtime implementations are used. The derived modelling rules are implemented as extension to the TNCES-Workbench and allow an update of the formal model, if the control application or the mapping changes.
Using a library of modules describing the discrete behaviour of often used mechanical components and parametrizing the temporal behaviour of them, the formal model of the uncontrolled plant is created fast and flexible. Nonetheless, the model of the workpiece has to be developed for each plant manually.

After connecting both models in closed loop a reachability analysis is done and it is shown, how a control engineer can examine in a graphical manner the system behaviour for all possible cases. Therefore, the TNCES-Workbench provides several graph layout algorithms from the *Graph Visualization Software* as well as the possibility to draw interesting trajectories as Gantt-Chart. Since the model of the plant incorporates all sensors and actuators, the state of the corresponding places can be included into the Gantt-Chart.

Even if the control engineer has no deep knowledge about the used formal model, this systematic and tool supported way ensures the possibility to analyse the system behaviour in any case, if he changes the control application or remaps it. This will reduce downtimes during production changes and new plant can be brought faster into service.